# STATISTICS for QUALITY CONTROL

## Daniel Jackson

Industrial Press, Inc.

Industrial Press, Inc.
32 Haviland Street, Unit 2C
South Norwalk, Connecticut 06854
Tel: 212-889-6330, Toll-Free: 888-528-7852
Web Site: industrialpress.com, ebooks.industrialpress.com
E-mail: info@industrialpress.com

Copyright © 2015 by Industrial Press, Inc.
Printed in the United States of America.
All rights reserved.
This book, or any parts thereof, may not be reproduced,
stored in a retrieval system, or transmitted in any form
without the permission of the publisher.

ISBN 978-0-8311-3517-1

Sponsoring Editor: Jim Dodd
Developmental Editor: Robert Weinstein
Interior Text and Cover Designer: Janet Romano-Murray

This book is intended as a guide. The reader is solely responsible for ensuring full compliance with all local, state, national, or regional legislation with respect to purchasing, procurement, and maintenance. Neither the publisher nor the author shall be responsible for the reader's non-compliance with any legal requirements. Any similarities to specific vehicle projects are purely coincidental. No purchasing project information is used which was not previously in the public domain.

MINITAB® and all other trademarks and logos for the company's products and services are the exclusive property of Minitab Inc. All other marks referenced remain the property of their respective owners. See minitab.com for more information.

Excel® is a registered trademark of Microsoft Corporation in the United States and other countries.

*In Loving Memory*
*of*
*Carlton L. Jackson*

# Table of Contents

Introduction ................................................................................. vii

A Note from the Author ................................................................ xiii

**ONE**      **Numbers, Measurements, and Scales** ............................... 1
             The Numbering System ................................................. 3
             Units of Measure ........................................................... 5
             Scales ............................................................................. 13

**TWO**      **Sampling and Organization of Data** ................................ 17
             Sampling ........................................................................ 18
             Organization of Data ..................................................... 23

**THREE**    **Basic Statistics for Discrete Distributions** ..................... 43
             Classical Probability ..................................................... 43
             Relative Probability ...................................................... 44
             Subjective Probability .................................................. 46
             Probability Variations ................................................... 47
             Counting Possible Outcomes and Events .................... 51
             Discrete Distributions ................................................... 53

**FOUR**     **Basic Statistics for Continuous Distributions** .................. 63
             Measures of Central Tendency .................................... 64
             Measures of Dispersion ............................................... 70
             The Empirical Rule ....................................................... 76
             The Standard Normal Z-Distribution ............................ 78
             Abnormally Shaped Distributions ................................ 82

**FIVE**      **SPC for Attribute Measures** ............................................. 87
             Individual Data versus Subgroup Data ........................ 88
             Common Control Charts for Attribute Measures ........ 89

**SIX**       **SPC for Continuous Measures** ........................................ 103
             Control Charts for Variable Data ................................. 104

**SEVEN**    **Control Chart Analysis** ..................................................... 119
             Variable Subgroup Sizes ............................................. 119
             Process Analysis ......................................................... 123

Table of Contents

|  |  |  |
|---|---|---|
|  | Process Control Limits, Tolerance Limits, and Specifications.... | 126 |
|  | Process Capability | 128 |
|  | Individuals Charts | 133 |

**EIGHT  Acceptance Sampling and Inspection ............................. 137**
- Sampling ..................................................................... 137
- Types of Plans ............................................................ 140
- Producer Interests versus Consumer Interests ........................ 147
- The OC Curve for a Specified Single Sampling Plan ................ 148
- OC Curve Properties .................................................... 151
- Other Common Acceptance Statistics ................................ 152
- Automated 100% Inspection ......................................... 154

**NINE  Basic Inferential Applications ........................................ 157**
- Confidence Intervals .................................................... 158
- Experimental Tests ...................................................... 162

**TEN  Additional Uses of Statistics in Industry ......................... 179**
- Product and System Design .......................................... 179
- DOE ........................................................................... 183
- Market Analysis .......................................................... 187
- System Operation ....................................................... 191
- Other Applications ...................................................... 194

**ELEVEN  Quality Awards, Standards, and Quality Management ....... 197**
- Awards ...................................................................... 198
- Standards and Societies ............................................... 200
- Six Sigma .................................................................. 202
- Lean Six Sigma .......................................................... 206

**Appendix A  Individual and Cumulative Terms of the
Poisson Distribution ........................................................ 211**

**Appendix B  Areas Under the Normal Distribution .............................. 223**

**Appendix C  Critical Values of the t-Distribution ................................ 227**

**Appendix D  Critical Values of the F-Distribution ............................... 230**

**Appendix E  Data Sets for *Statistics for Quality Control* ..................... 234**

**INDEX ........................................................................................ 235**

# Introduction

Probably because of Adam Smith and *The Wealth of Nations* (1776), and later Charles Babbage and *On the Economy of Machines and Manufacturing* (1832), philosophies emerged throughout the 19th century changing the way industrialists view the resource of labor and how industry should operate. These philosophies, along with key technological and scientific developments, came together in the second half of the century that defined the industrial revolution (in the United States much earlier than in Europe). These key technological developments included access to energy, transportation, and standardization. Along with advances in science, and particularly math, this period marked an incredible increase in mechanical thinking, inventions, and innovation.

The mechanical renaissance of the industrial age marked many important milestones. Because of the steam engine, coal mines became more productive, allowing timely access to energy. Using this energy, the steam engine not only powered industrial equipment, but also brought a remarkable technological development to transportation. Within a few years, trains, boats, carriages, and even early automobiles were powered with the steam engines that fulfilled yet another condition for the industrial age to flourish — the cheap and effective delivery of raw materials.

An early invention that helped launch the industrial age was Eli Whitney's cotton gin, which paved the way for a sudden glut of cotton in the United States. With the influx of cheap cotton, abundant energy, and a capable transportation system, the textile industries flourished. This growth gave the United States the economic stimulus needed to compete in a variety of industries worldwide. However, from a perspective more pertinent to industry, Whitney went on to invent and develop manufacturing processes leading to the standardization of parts. Initially used in rifle production, standardization was quickly adapted to industry in general. It also transformed the operative worker from a skilled craftsman to a semi-skilled laborer with the ability to produce parts virtually identical to another by using specific methods, materials, and machines.

Matching the technological advances in machinery and precision were the operations strategies of labor management and standardization, the ready supply of fuel and transportation, and the continuing exploration of math and science. With all these factors in place, the industrial age was poised to grow exponentially for the remainder of the 19th century, accelerating into the 20th century.

## Early 20th Century Developments

The word *industry* comes from the Latin word *industria*, meaning diligence at work. The industrial age certainly presented an environment living up to the term. Industrialists and laborers worked diligently with "elbow grease" and the "sweat of their brow." It was a busy, exciting time in our world's history. The idea of mass producing a product had never been put to such a practical use. What was lacking was efficiency, attention to the cost of wasting materials, quality, and competitive improvements in design.

Frederick Taylor contended workers typically were afraid that if they worked as hard as they could they would work themselves out of a job. They also reasoned, with their wage pay, they would receive the same pay regardless of the work performed. He termed this phenomenon *soldiering*. Taylor was convinced workers should toil efficiently through a method determined by science rather than rule-of-thumb. They should then receive pay incentives for established production quotas. Management should accommodate this scientific method and, through a spirit of cooperation with the laborers, efficiency would improve. *Taylorism*, as it became known in a derogatory sense, did greatly increase efficiency, but at the same time, henceforth, caused layoffs. He saw *Scientific Management* as a means to increase the United States' national efficiency as urged by then president Theodore Roosevelt. Laborers and managers alike were vehemently opposed to Taylor's reforms because it represented such a marked change from tradition. In 1911 Taylor published *Principles of Scientific Management*, later becoming the basis for Operations Management, of which industrial statistics is a part.

Frank and Lillian Gilbreth, contemporaries to Taylor, worked as a team to increase the efficiency in the motions required for operation tasks. Their research was mainly studying workers during production, typically by film, and suggesting different motions to increase production. The Gilbreth's contributions were received slowly, as were Taylor's. However, Frank Gilbreth documented their work in *Motion Study* in 1911, and then in 1919 the Gilbreths jointly published *Applied Motion Study*.

The effect Scientific Management had on operations was staggering. Henry Ford implemented many of these principles in his Model-T plant. These improvements, most notably in the assembly line, may not have been realized without the contributions of Taylor and the Gilbreths; they led to the production of over 15 million cars between 1909 and 1923. These numbers alone were a wake-up call to all manufacturers. To society, it was something new and amazing that changed the lives of the lower class (the majority of the population at the time). With the improvements in mass production, a large segment of the population had affordable access to products normally used only by the wealthy.

Perhaps the philosophies and principles of Scientific Management were too new during World War I, but little of this science nor much of statistics was utilized in the war effort. This is in direct contrast to World War II. The time between the wars shows a continued interest in Scientific Management, its development into Operations Management, and the increase in the use of statistic in industrial operations.

To facilitate a study based on statistics, accuracy of precision measurement was a must. With the micrometer, the Vernier caliper, and now gauge blocks, the ability not only to manufacture an item to an accurate unit of measurement, but also to view the minutest increment became possible. It is the key to any industrial statistical study. This precision allows scientists to discern the slightest variance in a product. In turn, it allows the collection of those data, making statistical analysis possible.

In Europe, one early industrial statistics pioneer was William Sealey Gosset at the Guinness Brewery in Dublin. Although Gosset was initially posted at Guinness as a chemist in 1899, he quickly became involved in statistics as a tool for quality control. He briefly attended University College in London where he studied under Pearson, one of the founding fathers of the science of statistics. Gosset's most notable invention was the t-test disseminated under the pseudonym "Student" in 1907. The t-test was used for making tests when there were very few samples allowing him to control the level of yeast and, hence, quality in beer. In this setting, it was only a matter of time before statistics was introduced into other industrial applications. Manufacturing companies throughout Europe began hiring scientists to conduct research into ways to improve operations.

In the United States, similar trends occurred, only later. Most notable was the research activities at the Western Electric Company. Walter Shewhart joined Western Electric in 1918 as an engineer. By early 1924, he had developed the first control chart. This control chart was based on the fact that there are both chance causes and assignable causes of variation in the manufacturing process. If the chart was balanced due to chance causes, the process was under control. If not, the chart exposed assignable causes, drawing attention to a problem or "out-of-control" condition. Today these charts are commonly used worldwide and mark the beginning of what we call Statistical Process Control (SPC). Shewhart published two important books dealing directly with industrial statistics and quality: *Economic Control of Quality of Manufactured Product* in 1931 and *Statistical Method form the Viewpoint of Quality Control* in 1939.

Colleagues of Shewhart's at Western Electric, Bell Laboratories, included Harold Dodge, Harry Romig, and Joseph Juran. This team of scientists produced the most significant contributions to the use of statistics in industry. These included not only analytic developments such as the Dodge-Romig Tables (MIL-STD-105) for standardized sampling plans, but also managerial philosophies regarding quality. Regarding the latter, Shewhart's *Plan, Do, Study, and Act* cycle (PDSA cycle or Shewhart cycle) is commonly regarded as the beginning of the Total Quality Management (TQM) philosophy. These achievements were so notable in increasing productivity and quality at Bell Laboratories that they were initiated at several other manufacturing companies. Generally, results were the same and the word spread.

During WWII, each of these scientist's contributions became integral to the war effort. With so much of the labor force joining combat activities in Europe and Asia, a large void was left in manufacturing, at a time when manufacturing was a critical requirement for winning the war. With an accumulation of technological developments from the recent past, standardization of parts, scientific management, and now improvements in quality control and sampling plans, the results for the war effort were highly successful. Even prior to the Japanese bombing of Pearl Harbor in 1941, the U.S. military employed these principles to speed production of war-related products and their delivery to allies in Europe and Asia. At the factory, the workforce shifted from experienced males to inexperienced females. However, with these managerial and quality innovations, the inexperienced women actually increased productivity of these items, overwhelming the competition (the enemy) with a steady supply of quality warfare product.

As the normal workforce returned home at the end of the war, production actually returned a surprising result — quality fell. With the momentum built during the war, manufacturing settled back to the old way of operating by producing in volume to achieve acceptable product. Because the country was working, resources were plenty, and the economy was booming, little notice was given to the dangers of such a practice. As frustrating as this must have been to advocates of the new discipline of quality control, it set the stage for what shaped the second half of the century and beyond, and eventually changed the way the manufacturing world operated.

## Late 20th Century Developments to Present

Following WWII, Japan was devastated. They had a limited workforce and no natural resources; economic and political structures were crushed and uncertain. Moreover, their almost non-existent manufacturing produced product considered to be of the poorest quality in the world. The United States set about reconstructing the country from all social-economic levels. This included reconstruction of Japan's ability to sustain its own economy. The key was industrial enterprise.

In 1950 Edwards Deming took on the task of helping Japanese manufacturers increase quality and production. Deming, having worked for the U.S. Department of Agriculture and the Census Bureau, was a private statistical consultant and professor at New York University in the Graduate School of Business Administration. Over the next several years, Deming was successfully teaching these principles in Japan whereas in the United States they were dismissed. Japanese manufacturing quickly picked up pace improving the economy. As their economy grew, it stimulated the interest of other entrepreneurs and ventures. Some refer to this period as the Japanese economic renaissance. Deming continued to consult in Japan; by 1970, manufacturing was competing not only on a price tag level but also on a quality level. By 1980, the quality of Japanese products exceeded that of U.S. products — and they were less expensive. This was a tremendous wake-up call to manufacturing in the United States.

In 1980, NBC broadcast a White Paper report entitled "If Japan Can...Why Can't We?" American manufacturers sighted no discriminating techniques making Japanese manufacturing successful; however, they did identify Edward Deming and noted his statements regarding quality not coming from inspection. This was a turning point for American industry. From that time, American industry was more receptive of Deming's principles and U.S. industry began to improve. By the 1990s, American industry was well caught up in the quality issue. However, further issues lingered regarding quality at an affordable price.

Deming's principles were centered on the 14 points outlined in his 1982 book *Out of the Crisis*. Although his system required a thorough knowledge of statistics and the application of those statistics in quality control, many of the principles surrounded a philosophy of quality management. Deming asserted quality came from the CEO. If management was not involved in building quality into the design phase of the operation, quality and efficiency would not exist.

Deming wrote several books and articles and became known as the "profit of quality." He is a national hero in Japan and the United States and is the honored center of the coveted Deming Award.

Contemporaries with Deming include Crosby, Juran (mentioned previously), Ishikawa, Taguchi, and many more. Crosby concentrated on the cost of quality and asserted ultimately the cost of quality was free. He developed the *Zero Defects* concept and wrote several popular books on quality management, including his most notable *Quality Without Tears* in 1980. Juran, initially starting with the quality team at Western Electric, wrote *The Quality Control Handbook* in 1951. This book continues to serve as the backbone for industrial statistics and quality management. Later, Juran founded the Juran Institute. Ishikawa developed Cause and Effect Diagrams popularly known as *Fishbone Diagrams* and advocated the use of Quality Circles prevalent in industrial operations worldwide. Taguchi developed the Quality Loss Function and the Robust Design strategy among others. These scientists molded modern industry into efficient operations producing quality product through changes in management method, the articulation of scientific principles, and an underlying philosophy regarding quality.

By the end of the 20th century, industry had transformed from using somewhat primitive industrial age production methods into a continuously improving quality revolution. SPC and quality management philosophies were widely used, the American Society of Quality (ASQ) had grown to over 100,000 members, and the Deming and Baldrige awards were established commemorating efforts in quality. Now efforts are underway to ensure a standard quality throughout the world within a lean and continuously improving environment.

## Use of Industrial Statistics in Manufacturing

Industry is an all-encompassing word that is attached to everything humans produce. Traditionally, industry meant the manufacturing of a given tangible item. Currently, industry can mean any broad category of any endeavor or business that makes money within the social economic system worldwide. Are motion pictures an industry? Are hospitals and clinics industry? What about farming and agriculture, tourism and travel, medicine, television, music, and art? Of course, the answer to all of these is yes. Each of these, and many more, represent a dissemination or distribution to society for commercial purposes. All represent a production of a product or performance of a service.

They are, however, varied in the effect they have on the customer. As such, each requires assessment of quality focused around the product. How would a motion picture be assessed for quality? Would it simply be box-office revenue? If so, the industry could single out the best movies using revenue data and could duplicate the efforts producing the next best movie. It is not that simple. With all quantitative measures, one must consider qualitative measures. In other words, quality does not come from numbers alone. As with the movie example, true quality comes from a more affective study on what stimulated the pleasure, intrigue, emotion, and experience from watching the movie. The quantitative measure aids the overall decision.

Quality in industry comes from a philosophy of producing a product or performing a service giving a set level of satisfaction to the customer. Industry can produce 800,000 parts, then by sifting through them, find the best 100 to sell to the customer. The results would be great customer satisfaction. However, it would be better for this industry to make 100 flawless parts to begin with. The customer satisfaction would remain the same, but the cost would be less to make and less for the customer to buy. To do this, quality must start with the production system and service design. Once quality is built into the system, the resulting product is of high quality supplied to the customer at a lower price.

# A Note from the Author

Statistics in quality are used to gather information regarding customer satisfaction, making decisions regarding the design and redesign of the process, and revealing the process during operation. Depending on what is measured, these statistics can determine, infer, or suggest estimates of effectiveness, dimension, use, and satisfaction.

This text concentrates on the use of statistics for SPC (Statistical Process Control) in a manufacturing operations environment in industry. But understand, although the text focuses primarily on quality control, statistics may also apply to many different functions. These functions include market research, research and development, and financial scenarios dealing with a variety of system design and operations perspectives. In each case, statistics help lead the organization to a better or improved design founded on the voice of the customer.

Chapter One introduces the different types of numbers, measures, and scales that have emerged throughout the years. Chapter Two explains how to collect, organize, and treat data using tables and graphs. Chapter Three discusses basic statistics for discrete and attribute type data whereas Chapter Four looks at basic statistics for continuous type data.

Next, the text concentrates on SPC, first for attributes (Chapter Five) and then for variables (Chapter Six). Chapter Seven explains how to analyze SPC charts and suggests means to improve quality. Chapter Eight continues the discussion on quality, but introduces acceptance sampling and a means to assure quality in industrial operations.

Chapter Nine introduces the basic inferential statistics used to perform experimental studies. In turn, Chapter Ten explores several other types and applications of statistics found in industry. Finally, Chapter Eleven focuses on the management of statistics to improve and maintain quality.

In each of these chapters, follow along with the examples and use the provided data sets to explore, change, study, and learn the practice of statistics in industry. Data sets for several of the end-of-chapter problems, as well as for some of the on-going examples (and accompanying figures) are provided online for use with both Minitab® Statistical Software and Excel®. See Appendix E for additional information.

As Deming said, quality cannot be inspected into product. To obtain a quality product in manufacturing, management must view quality as a system-wide requirement. In order to improve quality, operational staff and management must monitor and improve the process of making the product. As Deming and many of his contemporaries have said, such improvements depend on a thorough knowledge of statistics.

I would like to acknowledge my graduate assistant Joleen Byerline for her edits, photos, time spent seeking permissions, and generally for contributing to this project. I would also like to thank the team at Industrial Press in supporting and having confidence in me throughout this project including Robert Weinstein, Janet

Romano-Murray, and of course Jim Dodd along with all the others behind the scenes putting this book together. Finally, I would like to thank my lovely wife Ling and two wonderful little boys Oliver and Cranley for having unending patience with me while involved in this project.

Daniel D. Jackson
*Western Kentucky University, Bowling Green, KY*
*March/2015*

# ONE
## Numbers, Measurements, and Scales

There is no such thing as an exact measure. In theory, measurements are only estimates. For practical purposes, however, these estimates are completely accurate to serve their purpose. Suppose a manufacturer of crankshafts for car engines wants to measure the diameter of one of the journals of the crankshaft. If the surface of the journal is examined under a powerful enough microscope, they would see the surface is uneven. Even if the surface was honed and polished with the most sophisticated technology available, it would still show some level of imperfection at some level of magnification.

In this case, how do they take a measurement? Do they average the imperfect surface? Or do they settle on the outermost surface portions, in which case, if a mechanical measurement instrument is used, how much pressure do they exert to take the measure? Once they have made contact with the surface of the journal with the measuring instrument, they have already indented that portion of the surface — regardless of how hard the metal is. If they increase the pressure, they will be measuring below the outermost dimension.

For practical applications, this microscopic depression of material would mean nothing to the function of the journal, but it does illustrate the fact the measure is an estimate. The true question asks how precise the measuring instrument is in accommodating dimensions of a practical application.

Suppose a micrometer is the instrument used. The micrometer is truly an example of Plato's assertion that "necessity is the mother of all inventions." It (the invention) was developed in the mid 19th century (adapted from the earlier telescope version used in astronomy) to improve the accuracy of length measurements for purposes of precision in machining (the necessity). Had machining already been precise enough, the micrometer would not have been invented. The micrometer can easily measure to the thousandth of an inch (0.001 in); those with an attached Vernier scale can get even closer. Although this measure is plenty close for the crankshaft journal and many other applications, it is still an estimate of the actual distance. Even with dial indicators,

**2** Chapter One

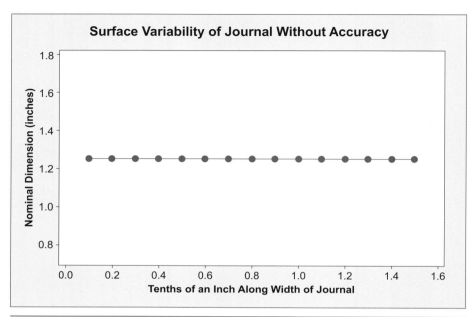

FIGURE 1.1   Poor accuracy

gauge blocks (a.k.a. jo-blocks), infrared, and laser measurements, it is only an estimate of the variability that exists along the surface of that journal.

Precision in the measuring instrument becomes critical when trying to reveal the variance in any measurable entity. Imagine trying to measure the variability of the journal surface on the crankshaft with a handheld one-foot ruler. Take a diameter measurement every tenth of an inch along the length of the journal and plot that number on a graph. The result is a perfectly straight line, as indicated in Figure 1.1. Do the same with a dial indicator (easily accurate to 0.0001 in) and the graphed line will start to show a variance in the measurement by moving up and down along the length, as shown in Figure 1.2. The more sophisticated measuring instrument reveals a closer measure than the ruler.

Accuracy is another consideration. Although the terms *accuracy* and *precision* are used interchangeably, there is a difference. Accuracy asserts the measure uses a scale appropriate for the characteristic being measured whereas precision determines how closely that scale measures the characteristic. The example just given was a distance, or length measure. So the scale (an inch ruler) was representative of distance but was not precise enough to reveal thousandths of an inch. Therefore, that scale was neither precise nor accurate because the measure was not a true representation of a thousandth of an inch. Other measures such as weight, volume, volts, temperature, parts-per-million, translucency, or dielectric strength require measurement instruments representative to those measures. One wouldn't measure temperature with a tape measure. In addition, each of these instruments must be capable of revealing a measure close

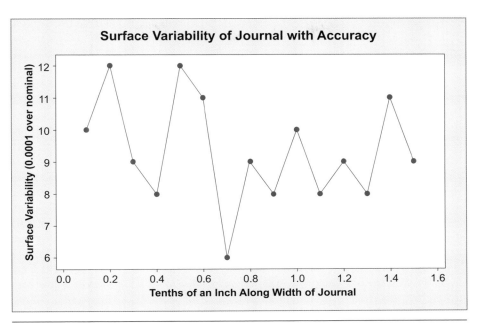

FIGURE 1.2  Adequate accuracy

enough to expose variation. Some measures become quite complicated and often times require a relative scale to describe them. These scales are considered subjective and are covered in more detail later in this chapter.

The accuracy of any of these measuring instruments depends on their ability to truly measure what the scale and precision purport to measure. Statistics is such a measurement system and, in many ways, a scale that measures and estimates the variance in a set of data. There are both precision and accuracy considerations. Don't let the term *estimate* fool you. By estimating and providing a probability of accuracy and precision, as most statistics do, statistics can come closer to the true theoretical measure than the measuring instrument itself.

## The Numbering System

Contrary to Leopold Kroenecker's statement "God made the integers; all else is the creation of man," mathematics was in full form prior to humans, earth, and the universe. There have never been any inventions in mathematics — only discoveries and attempts to explain a natural science. This was no trivial undertaking during man's time on earth. Followers of the Greek mathematician Pythagoras even murdered Hippasus because he discovered the existence of irrational numbers; mathematicians still struggle with inconsistencies presented in our cumulative explanation of this science. Given the mathematics used in statistics and the statistics used in industry, this theoretical realm of mathematics rarely presents problems in its practical use. Still, to have a clear understanding of scales and measures, there must first be a clear understanding of numbers.

Numbers are different than numerals. Numbers are almost an instinctive concept to humans. Undoubtedly, because humans have five fingers on each hand and five toes on each foot, counting was initially relative to these extremities. Perhaps in trade and commerce, humans even conceptualized counting (called natural numbers) by having two hands and two feet (20) of sheep or cattle. Then the idea of representing this concept with a symbol emerged.

The earliest evidence of this was probably the act of notching bones in Africa some 37,000 years ago. These "tally sticks" may represent a calendar or accumulation of game or crops. Written numbers or numerals began much later and developed alongside that of writing. The Sumerians (base 60), then Egyptians (base 10), Indians (eastern, unless otherwise noted), and Chinese developed symbols for numbers starting around 7000 years ago. Because of proximity, these systems may have borne some influence on each other. The Mayans, however, being detached from the Indo-European land mass, developed their own system using base 20. The Indians were probably the first to develop a symbol for zero, which was somewhat of a new concept at that time. Today, whole numbers include all positive and negative numbers, and zero. Many systems left out the zero, creating confusion between 492 and 4092, for example.

Since then, humans have complicated numbers with various classifications. The following definitions will help. It is a partial list that includes the kinds of numbers relevant to statistics used in industry. Understand that there is much disagreement in the science of mathematics in how best to define these terms.

**Counting numbers or natural numbers.** These are 1, 2, 3, 4, 5…to infinity. Zero is not included because zero is a non-countable number; when people first began to count the very concept of zero was nonexistent.

**Integers.** The set of natural numbers and zero.

**Whole numbers.** The full set of numbers including natural numbers, zero, and negative numbers (no fractions) from negative infinity to positive infinity.

**Rational numbers and Irrational numbers.** A rational number is a number between two whole numbers giving a precise value. For example, 1/2 = 0.5. On the other hand, 2/3 = 0.66666… is an irrational number because it never ends. In practical use, this number terminates with a 7 at a given placement value, for example, 0.667 or 0.66667.

**Cardinal numbers.** The names of numbers in any given language. In English, these are one, two, three, and so on.

**Ordinate numbers.** These are the names give to numbers representing a sequential order, such as first, second, third, fourth, and so on.

**Nominal number.** Nominal numbers have no mathematical value. They are simply numbers given to represent a certain item, entity, group, classification, or category. Examples include your social security number, zip code, or telephone number. In statistics, nominal numbers are called categorical data. For example, all items that are blue may be designated as one whereas all items that are red may be designated as two, or as in industry, zero means *off* and one means *on*. In some cases where there is no order, serial numbers are considered nominal. However, if the serial number represents a position in a sequence, it becomes an ordinate number. Nominal numbers are used extensively in statistics when labeling attribute type data, which will be explained in Chapter Three.

**Intervals.** Numbers that are meaningful when compared to other numbers. These numbers are not, however, meaningful when compared to zero. Zero may simply be another arbitrary point on the scale. Take common temperature scales, for example. The Fahrenheit scale set zero at the temperature ocean water freezes. Therefore, 10 degrees Fahrenheit is not half the temperature of 20 degrees Fahrenheit. Even at zero degrees Fahrenheit, the amount of heat is substantial. Fresh water boils at 212 degrees Fahrenheit, but it is not 212 times hotter than water at 1 degree Fahrenheit. It is true however that 10 degrees Fahrenheit contains the same amount of heat whether it comes from −10 to −20 degrees Fahrenheit or 90 to 100 degrees Fahrenheit. This is the advantage of the interval — equal amounts through the scale have the same value.

**Ratios.** A ratio represents the relationship of one number to another. Unlike the interval, the ratio is a number on a scale where zero *is* meaningful. The number of defects in a manufacturing line is one example. Zero defects per number of produced items means the number of defects cannot go lower. Length compared to width, weight to volume, strength to weight, and mixtures such as 1 part rice to 3 parts water are examples.

## Units of Measure

Most units of measure are manmade. For example, length is determined through comparison to another distance. This other distance could be anything — the cubit was based on the distance from the elbow to the tip of the middle finger. As scales developed through history, the measures became more standard. Still, homemade measures are often used today. When hanging two or more pictures on a wall, how many people fix a position on the first picture and compare it to their own eye level or chin level? This level then becomes the reference height for hanging the next picture.

Over time, measures became standardized so that information could be transferred farther than a short distance down the wall or hallway. Standardized scales are essential in statistics. Analyses might be achieved without them, but interpretation by the rest of the world is difficult. For example, weight data on steel rods could be collected using a beam scale. Place the steel rod on one side and use 1/4" grade 8 flat washers on the other side. Keep adding washers until the scale is level; then count how many washers. Every statistical analysis needed can be computed using these data and the meaningfulness of the results will be sound. However, by converting the washers into pounds, ounces, grams, or kilograms, this information becomes more easily transferred and understood by the profession around the world. It becomes standardized without changing the outcome of the statistical analysis.

Let's look at just a few of the most common standard measures. The list is partial because there are literally thousands of standardized measures.

## Length

There are two common standards in measuring length: the British standard scale and the metric scale. The foot was developed just as one would guess. However, feet vary between individuals, so several different foot measures evolved. The foot was standardized when the Normans came to England, the 12-inch foot was made official by King Henry I less than 100 years later. The inch too was representative of a human part — the width of the thumb. It progressed as the foot did and was standardized as 12 per foot. The yard was the length between the nose and the outstretched arm. It was later standardized to equal three feet.

Today, the standard scale can be fractional or decimal. In fractional, an inch can be taken in half, then 4ths, then 8ths, 16ths, and further. Figure 1.3 shows a typical partition of a standard scale using fractions. The largest partition is the half, which is then split into quarters. The fractions are counted as 1/4, 1/2, 3/4, and a full inch. The

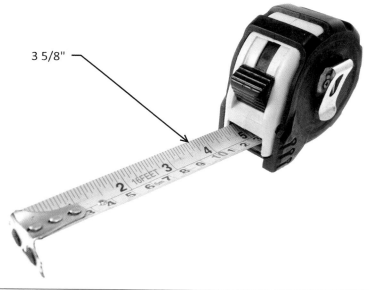

FIGURE 1.3   Inch scale

quarters can be split into eighths and then sixteenths. The arrow shown in Figure 1.3 indicates 3 and 5/8ths inches.

Decimal scales are similar but can be written using decimal place value format rather than fractions. The mile was introduced first by the Romans; it was defined as 1000 paces of a Roman legion. In most of England, however, this distance was thought of as 8 furlongs because a furlong was truly an English measure (representing a plowed furrow of 660 feet in agriculture), making a mile equal to 5280 feet.

The metric scale is available only in the decimal form. It was developed by the French Academy of Sciences commissioned by the National Assembly of France in 1790. One meter equals one ten-millionth of the straight line distance between the North Pole (geographic) and the equator. Most of the world (including England) uses the metric system to some extent. The United States is one of the last remaining countries that have not officially changed to the metric system.

Both standard and metric scales are capable of significant accuracy and precision, depending on the instrument used to take the measure. The scale itself is not determined by the measurement. Accuracy can vary between miles or kilometers, from nominal sizes of one or two inches or even centimeters to highly accurate readings in the millionths of either standard or metric scales. Just as one can measure the distance between New York to Beijing, one can determine the width of an atom. For example, a nanometer (nm) is $10^{-9}$ or 0.0000000001 of a meter.

Similar to length are area and volume. Common area measures include square inch, square foot, square mile, acre (0.0015625 square miles), square meter, square kilometer, and hectare (10,000 square meters). Volume and capacity includes cubic measures of most of what was included above for area. In many volume measurements, however, volume must be defined as liquid or dry. Pints, quarts, gallons, and liters almost always refer to liquid volume measures, whereas ounces and cups (proportions of quarts, gallons, etc.) frequently measure dry volume such as flour or rice.

## Weight

As with length, commonly used weight measures include British standard and metric systems. Uncommon weight measures are many and varied throughout the world. In China, for example, (prior to the 20th century) the jin equaled 604.79g. Now, in a conformance to international standards, this ancient measure equals 500g. All measures, including weight, were simultaneously developed worldwide. Through pressures and circumstances that go beyond explanation in this text, many have disappeared or were changed and absorbed into the modern international standard (usually metric).

Even the British standard system has made several measures obsolete, including the very definition of the pound. Today, the pound is standardized as an avoirdupois pound as opposed to the troy, merchant, apothecary, tower, or London pound. Because of these many different systems throughout history, there exists an ambiguity among measures. For example, the imperial ton, also called the long ton, is 2240 pounds; the short ton (used in the United States) is 2000 pounds; and the metric ton (or tonne) is 1000 kilograms or 2200 pounds (approximate).

Tons are divided into pounds, pounds into ounces, and ounces into grains. In the metric system, the gram is the primary weight unit. One kilogram (1000 grams) equals 2.204622621849 pounds.

## Electrical

Common electrical measures include voltage, amperes, ohms, and wattage. Volts are the potential power available. Think of a water tower as an analogy. The amount of power from the water pressure in the tank (the weight of the water) can be thought of as the volts. Most electrical applications require a certain amount of voltage to operate. Computer manufacturers are interested in what variance there is in the voltage available at the power supply and transformer. Tests would be performed on each transformer to determine if it met minimal voltage for the computer to operate.

Power available is not the only consideration. Even if power is available, the delivery rate may be too slow because of resistance. This is a measure of amperage. Again, thinking of the water tower, regardless of how much water is available, how much water can flow through a given hole in the bottom of the tank? Given a substantial amount of water, a small hole creates a fast spray of water, but the volume of water delivered is minimal. A large hole creates a slow gush delivering a large amount of water volume. The volume of water moving through the hole (dependent on pressure and size of hole) is the amperage. The size of the hole is resistance. So, an electric motor in a washing machine needs the correct amount of voltage to activate the system, the availability of enough amperage to keep the motor running, and the proper amount of resistance to control the current to keep it from burning up.

These measures are read from meters. A typical multi-meter can detect voltage, amperage (current), and ohms (resistance). One of the most common measures in electricity is continuity. This measure is a "yes or no" measure. It either has electricity going through it or it does not. If antique automobile renovators are tinkering with the tail lights on an old car, they may want to see which wire from the switch goes to which light. Putting the multi-meter in the ohms reading mode and closing the circuit at each end of the wire can detect total resistance (no connection) or no resistance at all (the same wire).

Wattage is a calculated value of volts times amperes. It represents the total amount of electrical energy and can be described using other power units such as Joules per second, Newton meters per second, and even horsepower. Several other measures dealing mostly with electronics include Coulomb, Farad, Weber, Tesla, Henry, and Siemens. The definition for each of these goes beyond the scope of this text.

## Light

Light can be measured both directly and indirectly depending on how a material behaves when exposed to light. Intensity, or brightness, is the most basic direct measure. Luminance is measured in candela per square meter, historically originating from the light emitted from one candle. This unit is sometimes referred to as a "nit". Wavelength or frequency of the wavelength determines the color. If the desired characteristic of a manufactured item is green, for example, measuring wavelength from reflected light off of this object can determine that. For green, the measure is approximately 500 to 565 nanometers.

Different materials display various properties when exposed to light; these include translucency, opacity, reflectivity, and refractivity. These properties can deliver an indirect measure of light and be important considerations of the various materials as affected by light. A transparent material — such as glass, some plastics, and water — will allow light to pass through. On the other end of this scale is opacity (not allowing light to pass through). In-between the two is translucency.

Typically measuring devices for light are sophisticated instruments and can be quite expensive. Consider, however, measuring translucency in plastic film. One inexpensive method is to make a homemade instrument with a flashlight, a film-holding fixture, and a photovoltaic cell (Figure 1.4a). Position the flashlight a constant distance from the photovoltaic cell. Shroud the flashlight where it adjoins the holding fixture so light cannot escape. Place a plastic specimen into the holder and read the voltage that the photovoltaic cell generates (Figure 1.4b). This process generates ratio data proportionally to the intensity of the light moving through the plastic. Although it is not a direct measure, it may be used for meaningful statistical analyses regarding that material. The scale is created by assuming 0.0 volts for a known opaque film, and measuring the voltage with no film installed (completely transparent).

Reflection and refraction are also open to clever ways of measuring. Reflection is the amount of light that is bounced off a surface. Refraction measures how a material changes the speed of light and, as a result, the light appears to bend through that material.

## Chemical

Chemical compounds and elements are usually measured in parts per million (ppm). The International System of Measurements (SI) uses the mole as the base unit that

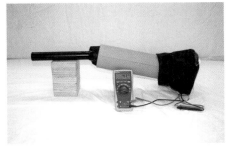

Figures 1.4a and 1.4b: Homemade measure of translucency instrument components
*(Photos provided by Marietta Byerline)*

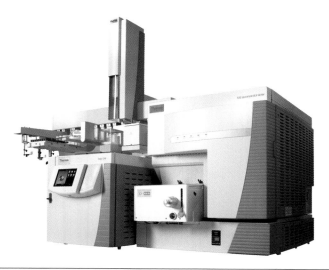

FIGURE 1.5   XLS ultra trace–1310 mass spectrometer
*(Photo provided by Thermo Fisher Scientific)*

defines a system with as many particles as equal to 0.012 kilograms of carbon 12 atoms. One ppm equals one micromole. These units of course must be qualified by volume, mass, fluid, or atoms.

Generally ppm instrumentation is expensive, accomplished by purity meters, lasers passed through a substance with the amount of light measured on the other end, and spectrometers. Spectrometers are probably the most sophisticated and most useful (Figure 1.5). Spectrometer accuracy is greater than ppm and identifies a variety of elements, as opposed to a dedicated instrument that identifies only the element for which it was designed.

## Viscosity

Put simply, viscosity measures the speed at which a fluid travels. Molasses flows slowly but water flows fast. A common additional variable is temperature. For example, oil flows quickly at high temperatures and slowly at low temperatures. It was this reason that necessitated the development of multi-viscosity oil for automobiles. Prior to the use of multi-viscosity oil, motors would suffer damage because the oil would be too thick to move in extreme cold weather and then too thin to lubricate when hot. An SAE (Society of Automotive Engineers) number of 10W-30 (the "W" means winter) indicates how viscous the liquid is at 0 degrees Fahrenheit and the viscosity in a warm engine. As the oil heats up, the viscosity changes little because of the polymerization of specific additives in the oil.

Viscosity is measured by varying types of viscometers (Figure 1.6) that return a unit such as Pa•s or the Pascal-second. Some viscometers force fluid through a hole and measure the amount over time whereas others drop a ball through the fluid and

# Numbers, Measurements, and Scales    11

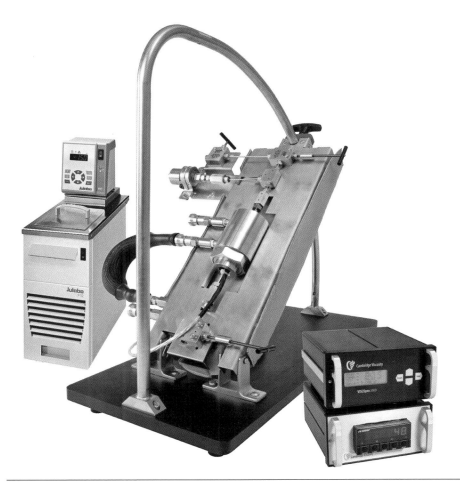

FIGURE 1.6  VISCOpro 2000 viscometer
*(Photo provided courtesy of Cambridge Viscosity, Inc., a division of PAC)*

measure the time it take the ball to travel a certain distance. You can also measure the length of time it takes for a bubble to travel a certain distance. Viscosity must take into account whether the fluid is liquid or gaseous.

## Heat

Heat is a basic measure of energy. In some ways, it can be quite easy to measure and, in other ways, quite complicated. The most common measure of heat is temperature. The standards for this measure include Fahrenheit and Celsius. The Fahrenheit scale (named after Daniel Gabriel Fahrenheit in the early 18th century) is based on the properties of sea water. Sea water freezes at zero degrees Fahrenheit and fresh water at

32 degrees Fahrenheit (originally set at 30 degrees Fahrenheit). Given this scale, water boils at 212 degrees Fahrenheit. Anders Celsius of the early 18th century developed a scale with 100 degrees between water freezing and boiling. This scale has become the most widely use standard in the world.

The simplest measure of temperature is with a mercury thermometer also developed by Fahrenheit. Other more sophisticated and hence exact measures include the use of piezoelectric devices. Thermocouples generate a voltage proportional to temperature and thermistors change resistance of incoming voltage proportional to temperature.

But measurement of heat doesn't end with temperature. Temperature is just one aspect of heat. Imagine a bathtub full of 110 degree Fahrenheit water. Then imagine a candle flame. The bathtub contains more heat but the candle flame is hotter. Other aspects of heat involve latent heat of fusion (the energy it takes to freeze a substance), latent heat of vaporization (the energy it takes to change from liquid to gas), and total heat or enthalpy (H). A hot day in Florida is defined differently than a hot day in the California desert. Because of humidity, the air in Florida contains more energy. Enthalpy represents the energy it takes to get the moisture (humidity) into gaseous form (latent heat of vaporization) and the actual temperature (sensible heat). In California, the total heat involves less humidity; therefore, less energy is used for latent heat of vaporization. Although the temperature is higher, the enthalpy is less. A loose measure of this is the "heat index" commonly reported during the weather segment of the local news.

Enthalpy, however, cannot be measured in temperature. Total heat requires a measure of energy (ability to do work), not just a single characteristic of that energy. Enthalpy is typically defined by internal energy (U) plus pressure (P) times volume (V) so that

$$H = U + PV$$

However, the more common measures of energy include joules (J) and British thermal units (Btus). The joule is the energy it takes to lift one kilogram a distance of 10 centimeters. Although this is a kinetic example, the amount of energy remains the same. Energy can just as easily be reported in calories (heat energy), watts (electrical energy), or horsepower (mechanical energy).

The Btu is truly a heat energy measure, as is the calorie. The calorie is based on the metric system and Btus on the British standard. One Btu can raise the temperature of one pound of water one degree Fahrenheit whereas one calorie can raise one kilogram of water one degree Celsius. Measures of energy are also useful in measuring the efficiency of energy conversions in thermodynamic systems. For example, an electricity power plant may use 100,000 Btus of input coal energy (coal is a solid chemical energy). The coal ignites in the furnace, releasing heat energy. In turn, the heat energy creates pressure (potential energy) by heating water in a closed vessel, the pressure moves a turbine (kinetic energy), and the turbine turns the generator. The generator produces 16,115 watts in one hour or 55,000 Btus of electrical energy, giving a conversion efficiency of 55%. Incidentally, this calculation suggests why electric vehicles recharged with electricity produced by a power plant actually use more energy than a vehicle using gasoline.

Another aspect of heat is the specific heat of a given material or, as it is sometimes called, heat capacity. Specific heat determines that material's ability to retain, release, or conduct heat energy. A good example is aluminum. Aluminum disperses heat quickly because it has a low specific heat compared to other metals. Because aluminum behaves this way, it is a good choice for internal combustion engine cylinder heads. An aluminum cylinder head keeps the operating temperature of the engine lower at the critical locations surrounding the combustion chambers. Aluminum's low heat capacity, however, makes it hard to weld. Typically, aluminum is welded with use of a tungsten inert gas (TIG) welder. The TIG welder uses a tungsten electrode (required for extreme high temperature) to produce an electric arc to the weld while shrouding the weld area with inert gas to keep the weld free from impurities.

On the other hand, liquid water holds heat very well, making it a good choice to circulate through a cylinder head. Water will absorb the heat, allowing the heat to be transferred to the radiator where it can be dispersed. Measuring specific heat is normally done by calculating the change in temperature of a host material (usually water) times its mass, divided by the change in temperature of the specimen material times its mass, or

$$C = (\Delta\text{temperature of water} \times \text{mass of water}) / (\Delta\text{temperature of material} \times \text{mass of material})$$

## Scales

Measures are data representing a single reading of a property. Scales, on the other hand, are the boundaries of those measures. For example, when measuring a distance, the reading is the number of inches or centimeters represented on the measuring instrument. However, the scale ranges from 0.0 to infinity. When addressing measures or units, it is almost impossible not to talk about scales. Apart from the description of unit measures, there are some general characteristics of scales that are necessary to discuss.

The first characteristic to consider is how objective the scale is. If the measure is inches, the logical scale is a standard instrument such as a tape, ruler, laser ruler, or yard stick. A properly taken reading with an instrument such as this gives an objective measure. On the other hand, if the measure is taken by non-standard instruments, such as pacing through the back yard to measure yards, or pressing your thumb to make an inch, it is subjective. Even more subjective is guessing.

The next characteristic to consider is that of quantitative versus qualitative. If the scale involves numbers and it is completely objective, the scale is considered a quantitative scale. Asking which truck is better, a Ford or a Chevy, is completely subjective, or qualitative. It depends also on what aspect of the truck is under consideration and how "better" is defined. However, if a statistician polled a large enough group of people who were truck owners, this qualitative question could be quantified with some objectivity.

Another subjective scale that has been quantified is the Scoville scale (named after Wilber Scoville and developed in 1912) for rating the hotness of peppers. The measurement comes from a panel of humans taking taste tests to determine when diluted hotness is no longer hot. Scoville quantified this into units representing the ratio of water to chili extract required to meet the criterion. An objective measure of hotness in chili peppers is the Gillett Method or high performance liquid chromatography (HPLC), which yields pungency units that are typically converted to Scoville units. A bell pepper measures 0.0 on the scale. A habanero measures 100,000 to 350,000, and pepper spray measures 2,000,000 to 5,000,000.

A qualitative scale attempts to measure specimens with expert subjectivity. For example, a wine taster determines which is the better wine using years of training and acquired knowledge from a variety of criteria. Supervisors rating worker performance in a factory also fit this type of scale. Qualitative scales can also be quantified just as any subjective scale, even though qualitative measures generally come from observation. For the supervisor, it is important to know that qualitative measures are just as important as quantitative. In fact, a qualitative measure can supersede a quantitative measure in many circumstances. The problem exists when only one or the other is used. Because quantitative measures are so easy to read and they give a precise level of measurement (although it may not be accurate), supervisors and managers tend to use them to make their decisions rather than investigating what might be the truth of the matter.

A final consideration is the nature of the measure and, hence, the use of an appropriate scale. The two most applicable types of measures pertinent to industrial statistics (particularly in statistical process control, or SPC, which is introduced in Chapter Five) are variables and attributes.

*Variables* are data that can change. They can take on values of increasing or decreasing quantities. There are two types of variables: continuous and discrete. A *continuous* variable represents an infinite number between units of measure. Distances, weight, proportion, electrical measures, and hardness are just a few examples of continuous variable measures. Can there be a 0.5 of the quantity measured, or for that matter a 0.529375? If so, it is a continuous variable scale and it determines the methods statisticians use to analyze them. For example, it would determine which control chart to use in an SPC study (see Chapter Six). As we'll see, the continuous variable requires an Xbar and R chart as opposed to a p-chart or an np-chart.

A *discrete* variable is one where data can change only from one whole number to another, for example, when counting people. Although the U.S. Census Bureau reported in 2005 that the average family had 3.26 people between the ages of 35 to 39 years old, it is simply impossible to have 0.26 of a person. There can be one person, two people, three people, and so on. People, automobiles, pencils, barrels, and all whole items that cannot be sold separately or split are examples of discrete variables. As can be seen from the Census Bureau example, discrete variable can utilize classic statistics as well. The problem is interpreting these statistics into something meaningful. The Poisson distribution covered in Chapter Five is a useful tool in analyzing this type of variable.

Both types of variables enable us to compute classic statistics such as mean and standard deviation in an attempt to describe the shape and behavior of the data — or what is referred to in statistics as the distribution.

*Attributes* are just what the name implies, a distinct quality of value. If market research indicates most people want to buy a blue t-shirt, then *blue* becomes that quality of value, or attribute. The t-shirt is either blue or it is not blue. It is not a question of how blue the t-shirt is, but a question only if it is blue or not. If the t-shirt is blue, it may be given the value of 1; if it is not blue it would then be given the value of 0.

Attribute data are used frequently for quality control in industry with go/no-go gauges (Figure 1.7). Suppose the desired quality of the diameter of a hole, such as the one shown in the figure testing a valve guide for a cylinder head, is a maximum of 0.223 inches. Even though 0.223 inches is a continuous variable measure, this hole might be tested by inserting a plug gauge for this particular dimension and tolerance minus the desired clearance between the shaft and the hole. If the plug gauge goes into the hole, the diameter is good. If the plug gauge cannot enter the hole, or is too loose, it is bad. Sometimes these plug gauges are color coded; green means go whereas red means no-go. The red side of the plug gauge is sized not to enter the hole. Good or bad, yes or no, on or off, 1 or 0, and one or the other are all examples of attribute measures. Attribute data are analyzed differently than variable data and are covered to a greater extent in Chapter Three.

FIGURE 1.7  Go/No-Go plug gauge
*(Photo provided by JIMS USA)*

# Chapter One Discussion

1. Why is there no such thing as an exact measurement?
2. Why is measurement technique important?
3. What is the difference between accuracy and precision?
4. Why is accuracy important?
5. Why is precision important?
6. Historically, how did numbers develop? How about zero?
7. Historically, how did scales develop? What are the commonly used scales worldwide?
8. Try to determine some of the types of scales and measures that were not mentioned in Chapter One.
9. How do Celsius and Fahrenheit differ in origin?
10. What is a variable?
11. Why is a standardized scale important in industry?

# Chapter One Problems

1. Give several examples of counting numbers.
2. Give several examples of whole numbers; explain the difference between whole numbers and counting numbers.
3. What is an integer?
4. Distinguish irrational from rational numbers. Provide examples of each.
5. What is the difference among cardinal, ordinate, and nominal numbers? Give examples.
6. Third, fourth, and fifth refers to what kind of number?
7. Describe an interval and give an example.
8. What type of numbering system uses a continuum of 1 through 5 where 1 is "strongly agree" and 5 is "strongly disagree"?
9. How are ratios used? Give some examples.
10. Make up a non-standard scale of your own. Take several measures and compare them to other objects to see if your scale is accurate and precise.
11. Explain and give examples of a continuous variable.
12. Explain and give examples of a discrete variable.
13. Explain and give examples of an attribute.

# TWO
## Sampling and Organization of Data

Loosely defined, statistics are simply a fancy and mathematical way of guessing. More specifically, statistics are a random and unbiased estimation of the characteristics of a quantitative distribution. They are used to describe, analyze, evaluate, and draw conclusions of a larger group (make inference) using only a small but appropriate part of that group. This larger group is called the *population* and the small part of that group is called the *sample*. Although *population* usually denotes the number of people in a certain area — for example, the world — in statistics it can represent any item defined as the total number of all like subjects that exist.

Consider the population of the United States (approximately 320 million people). How tall is the average American? To determine the answer using the entire population would be unrealistic. By the time everyone was measured, there would be a change in the population because of immigration, births, and deaths. Many previously measured people would be shorter because of old age or taller because of growth.

Now consider a population of rivets produced during one shift at a manufacturing plant. One quality measure of a rivet is the hardness. If the rivet is too hard, it will not compress together and hold the material it is designed to hold. If it is too soft, granular plates of the metal will not "work harden" into place when the rivet is compressed, and the system will fail. Realistically, measuring the hardness of 100,000 rivets in one shift is not possible. If the number is smaller, perhaps 800 rivets, it may be feasible to inspect all of them. This number is referred to as a sample rather than a population. If the data is collected correctly, this sample is a good representation of the population, mirroring the characteristics of that population. If 100% inspection is used, the resulting average is called a *population parameter*. This parameter is a real number defining the population, not estimating or representing it. However, 100% inspections are laborious, expensive, unreliable because of technician fatigue, and — most important — unnecessary. For destructive tests, 100% inspection is impossible (there would be no more product to sell).

In addition to finding an appropriate sample, you must organize the resulting data into a meaningful representation. Data, when first measured and collected (raw data), are gathered in no particular order. They must be placed in tables, graphs, or charts to reveal a relationship to one another and to facilitate the calculations necessary for describing distributions and making inferences regarding the study.

## Sampling

How do manufacturers determine if the rivets are acceptable? They use a representative sample of the population. The sample represents the larger population and the average becomes a statistic, rather than a population parameter. Using samples has important implications. First, the sample must be truly representative of the population; otherwise, the information gleaned from the study will be erroneous. To make the sample representative, some homework is required. Choose subjects that are pertinent. For example, measuring hardness of the rivets for purposes of improving the process require rivets from that process, not from a subcontractor. Second, data should be randomly collected from the sample. Scooping the top layer of rivets off the holding bin of one machine does not relate to an entire process using several machines. It won't even be a good sample from that one machine. To collect a random sample requires collecting a few rivets from each machine over a period of time.

### Sample Size

The sample must contain a sufficient number of subjects. This *sample size* is not necessarily proportionate to the population; it requires a minimum number to fully develop a representative distribution. Assuming the data are variable, if properly sampled, a sample of 1000 is as effective for a population of 15,000 as it is for a population of 2,000,000. Measures of central tendency start to decrease in accuracy with samples below 120, although accurate measures begin to appear even with samples of 30 to 40. These measures and the normal distribution are covered in detail in Chapter Four.

A more analytic approach to determining sample size given a certain population entails an examination of error. If a sample average is different than a population average, this difference is described as error. By setting the acceptable error, using the population parameter $\sigma$(sigma), and determining the confidence interval, statisticians can calculate the sample size needed. Determining sample size using these methods will not be detailed in this text. Conceptually, however, there are three criteria. The first criterion is error, which is set by the investigator and is represented in a plus or minus form such as +/– 5%. The second criterion is confidence, which is how sure the investigator wants to be in the final results of the study. (Remember that statistics is a science of estimation, not of exact determination). Confidence is usually represented as a plus or minus value from the mean. The third criterion is the degree to which the measure varies within the population. The nature of the population will largely determine this. For example, determine the number of people with brown eyes. The variability of this measure is quite different between Indonesia and the United States, even using a similar population size and the same sample size.

Similar approaches are involved for both variable and attribute data. Conveniently, several tables are available to help determine sample size. The most common are included in the ANSI/ASQ Z1.4 and ANSI/ASQ Z1.9 standards. Select tables, along with other useful tables, are included throughout this text.

Calculating sample size will be revisited in later chapters as the need arises. Oddly enough, the most common method of setting sample sizes in industry is by guesswork.

## Randomness

Even more important than obtaining an adequate sample size in a statistical study is that of collecting the data in a completely randomized manner. Without random data, the study loses validity. There is no purpose for a study without validity. A manufacturing plant that wants to purchase material from a new vendor is not likely to request 3 or 4 of this vendor's best samples. In contrast, the plant requires a representative sample taken randomly from the vendor's operations. A random sample represents an average quality with a certain measurable variance among subjects. Otherwise, the vender may send a sample from one machine or may handpick the best specimens from a lot. When the plant then starts utilizing the material in normal operations, a variance and average quality not previously measured emerges.

In some cases, obtaining a random sample is quite elusive. Suppose statisticians investigating the height of Americans want a random sample across the United States. They may be compelled to divide the area of the United States evenly into a grid. Next, they randomly collect an equal number of data from each of the grid points (Figure 2.1a). This approach represents a non-random selection because the investigators fail to divide the population evenly as it is concentrated throughout the country. The resulting data places more emphasis on the heights of people in the low population sections of the country (perhaps Midwesterners grow taller than Northeasterners).

The solution to this problem is to weight the high population centers proportionally. Collect more data from those areas, more fully representing a cross-section of the country (Figure 2.1b). Then from those areas, randomly sample the proportional number of people. An alternative is to determine one state with known demographics that mirror the rest of the country. Florida, for example, is known as this kind of state. The cost and logistics of such a study are drastically reduced and the results are within an acceptable deviation.

## Bias

One important aspect of any statistical study is reducing all sources of variance (the way a measure varies) until only natural variance (random variation) occurs. Any deviation from random sampling introduces variance that is unknown. But there are other sources of unwanted variance. Regardless of the source, unknown variance will alter the validity of the study. Sometimes, this source of variance, which is called *bias*,

## Chapter Two

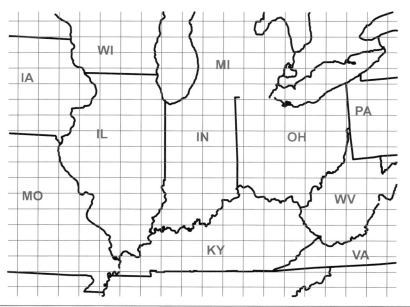

FIGURE 2.1a   Map evenly divided into a grid

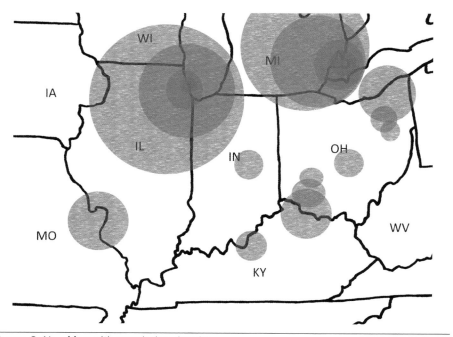

FIGURE 2.1b   Map with population density

is overlooked, ignored, or in extreme cases purposefully and unethically introduced. Just ask 500 people at a Democratic convention who will be president. Then ask the same question at a Republican convention and the answer will be different. This variance is probably the clearest form of bias, known as prejudice. It is quite obvious why a statistician would not and should not venture into such a study, but other forms of this bias are not so obvious.

This was the case in the famous 1936 U.S. presidential election in which pollster George Gallop predicted Roosevelt's victory over Landon. The most respected and longtime accurate predictor of elections at the time was the Literary Digest, a New York-based magazine. The Digest's strategy was simple. They polled 2,300,000 people using the phone book and automobile registration documents to locate them. Their resulting poll was overwhelmingly in favor of the Kansas Republican Landon. Gallop, on the other hand, polled only 50,000 people using random methods of selecting subjects and taking careful precautions against any source of bias.

Roosevelt won with the largest landslide in history. What was the Digest's mistake? In 1936, the voting majority didn't own a telephone, let alone an automobile. The Digest was polling only the affluent, which did not give a true representation of the way the country would vote. Hence, the Digest introduced a bias, which ultimately led to the incorrect conclusion.

Other bias creeps into studies (especially in industry), sometimes by laziness or by technicians having a "good enough" attitude. The purpose of the sample is to represent the population. If the sample is not carefully collected, the population is not represented and the study is a waste of time. Chances are the studies are for product and process quality purposes or for making strategic decisions that affect a company's success. Supervisors of these activities need to ensure all staff maintain a responsible attitude while sampling.

Sometimes, but with the same results, bias is caused by unethical practice. If lettuce suppliers place only the healthiest heads of lettuce on top of the lot so the grocer will buy, they are creating bias through an unethical practice. The sample is not only non-representative of the lot, it is also a misrepresented sample of the lot. Sometimes in industry, pressures to increase quality or quantity may lead workers to "fudge" the data in order to save their jobs, make themselves look good for promotion, or win funding of a government contract. Others may have a problem with the company or want to undermine a particular project, so they sabotage the data.

## Controlling Variables

Just as care should be taken in eliminating bias, data collectors should search out all sources of variance that would affect the sample. Some sources of variance may be pertinent to the study and would be introduced to the sample. Others, similar to bias, must be eliminated. It is essential data collectors understand and recognize which variables are needed and which are not for the study. They should have a basic understanding of statistics and know enough of the technology and science relating to the study to recognize and control these variables.

Suppose a company is commissioned by NASA to develop a shear pin for a rotating shaft on a satellite designed for a 10,000 mile geocentric orbit. What variables will be present under these working conditions? How will they test quality characteristics on this item? What variables need to be controlled?

This question is just as much a design question as it is a quality test question. Any variable included in the design must be present or simulated during the test. Some of the variables that might affect this product include mechanical stresses, chemical resistance, thermal stress, and exposure to high energy radiation. Apart from actual dimensions, the most obvious consideration is the mechanical stress. The shaft will turn with a maximum torque; therefore, the shear pin must be tested to withstand this.

So one variable is shear load. But how is the load introduced? Is it a slow steady increase or an abrupt instantaneous load? Perhaps the dynamics of the load need to be variable as well. A dynamic load affects fatigue of the pin and, hence, its life expectancy. Is there an exposure to chemicals? If so, this variable must be present in testing. Even lubricating compounds may behave differently in space than on Earth, making a difference in the performance of the part. The temperature at which the pin operates may be near absolute zero at times and extremely hot at others. The pin's mechanical properties change depending on heat or cold coupled with load. In addition to thermal energy, there is high energy radiation from the sun, such as ultra-violet radiation. A proper study testing this product should consider all of these variables, including the possible effect of operating this device in a vacuum.

Some variables may exist because of what is present in space. Others may exist because of what is absent. Variables caused by the presence of gravity, oxygen, nitrogen, water, and other earth elements are examples of what must be eliminated while testing this product. Otherwise, the environment in which it is tested is different than that of operation — the testing will not be representative of the product's normal working conditions. Statisticians must determine the procedure of measuring any quality characteristic with these variables in mind and then analyze the resulting data accordingly.

The actual collection of the data may itself introduce an unwanted source of variance. As such, those making the measurements should understand how critical their jobs are. For example, before turning operative workers loose on a dial caliper or a micrometer, make sure they have the necessary training on how to use those instruments. With other more sophisticated measurement instruments, a higher level of education might be required. In either case, technicians taking measurements should ensure the instrument is operating properly and is calibrated. Calibration procedures vary from instrument to instrument. Based on how strenuous the test is on the instrument, more frequent calibrations may be required along with what is called a *Gage R&R* test to check repeatability and reproducibility. Check that the scale, the measuring instrument, and the technique used to measure all meet accuracy requirements. While collecting data, develop a consistency in measuring technique. Consistency ensures a further reduction in variation that is not considered pertinent to the study or test. The variance left after all unwanted variables are identified and controlled is the natural variance that is part of any statistical study. This natural variance serves to describe how the distribution behaves; it allows researchers or quality personnel to make a description of the sample and informed decisions based on the distribution.

## Organization of Data

The resulting data from a random collection is just that — random. There is no apparent order in the data; trying to extract anything meaningful in its raw form is premature. A seasoned statistician can peruse the data and begin to get a feel for the distribution. However, any description, conclusion, or inference based on the data cannot be accomplished until they are organized into a readable form facilitating statistical calculations. The most common graphic organization of data comes in the form of frequency diagrams. Of these, the most widely used is a *histogram*. To construct a histogram, or any other frequency distribution, you must first organize the data into tabular form and then into a graphic representation. Table 2.1 depicts an example of 30 raw data. These data represent a certain chemical concentration in parts per million (ppm) found in pond water in close proximity to a manufacturing plant. What follows is the organization of these data.

### Tabulation

Tabulation is the first step in organizing raw data. Table 2.2 shows an ascending array where numbers increase from lowest to highest. Conversely, a descending array arranges numbers from highest to lowest (Table 2.3). Immediately the range becomes apparent as the data become organized. The range, which represents the difference between the highest and lowest (shown in bold), is a basic measure of dispersion covered in Chapter Four.

These data represent ppm of chloride in the water of a pond. Because this pond is in close proximity to a manufacturing plant using that chemical, the plant conducts small intermittent tests such as this to determine if the plant is within EPA guidelines regarding water contamination. Assuming the data are collected randomly, the 30 raw data can be arranged as they are in Tables 2.2 and 2.3 and then tallied.

A tally simply lists all possible readings; it indicates how many of each of those readings was present in the sample. Table 2.4 shows the tally of the data in Table 2.2 in ascending order. Once organized and tallied, these data can be used to calculate measures of central tendency such as mean, median, and mode, and measures of dispersion such as range and standard deviation. (These measures will be discussed in Chapter 4.) The data can be presented in a chart to graphically represent the way those data are behaving. To build this chart, reference the arrayed data (Table 2.2). Each reading is listed in the first column. Entries next to each reading, in each respective column, mark the number of occurrences. The tally itself begins to graphically represent the distribution.

TABLE 2.1  Raw PPM Data

| 266 | 278 | 255 | 255 | 278 | **237** |
|---|---|---|---|---|---|
| 255 | 258 | 243 | 241 | 253 | 243 |
| 247 | 269 | 260 | **279** | 252 | 268 |
| 271 | 264 | 273 | 256 | 274 | 257 |
| 262 | 267 | 250 | 275 | 255 | 263 |

## Chapter Two

TABLE 2.2  Ascending Array of PPM Data

| **237** | 250 | 255 | 260 | 267 | 274 |
|---|---|---|---|---|---|
| 241 | 252 | 255 | 262 | 268 | 275 |
| 243 | 253 | 256 | 263 | 269 | 278 |
| 243 | 255 | 257 | 264 | 271 | 278 |
| 247 | 255 | 258 | 266 | 273 | **279** |

TABLE 2.3  Descending Array of PPM Data

| **279** | 273 | 266 | 258 | 255 | 247 |
|---|---|---|---|---|---|
| 278 | 271 | 264 | 257 | 255 | 243 |
| 278 | 269 | 263 | 256 | 253 | 243 |
| 275 | 268 | 262 | 255 | 252 | 241 |
| 274 | 267 | 260 | 255 | 250 | **237** |

Suppose there are more than 30 data. Imagine how long Table 2.4 would be if there were more than those few data. The higher the n (number of samples in the distribution), the more cumbersome it becomes to read, tabulate, and tally the data. The process of calculating statistics also becomes lengthier. Grouping the data into intervals reduces the total amount of numbers in the tally, making it easier to read and calculate the required statistics.

## Grouped Data

The purpose of grouping is to reduce the number of data for both ease of display and statistical calculation. First divide the range plus one to find (preferably) an odd number. Then fill the resulting cells with readings within the cells' range, and determine pertinent points within the cell to calculate statistics and configure charts.

Figure 2.2 diagrams the terminology used in cell grouping. In this example, the cell interval — technically from midpoint to midpoint — is equal to 5. Some refer to the cell as the class or bin. The cell boundaries show the upper and lower limits of each cell boundary. In the case of cell 1, note the lower boundary is 9.5 and the upper boundary is 14.5. Increasing the decimal accuracy in the cell boundaries prevents confusion in which cell data fall. Again, 10, 11, 12, 13, and 14 fall into cell one. Cell 5 contains 30, 31, 32, 33, and 34. As there are 5 possible numbers available in each cell, the midpoint of the cell is the third possible number. In cell 5, the midpoint is 32. In cases where there is an even number of possible numbers in each cell, the midpoint is the average of those possible numbers. For example, suppose a cell contains 40, 41, 42, 43, 44, and 45. Following equation 2.1:

$$\bar{x} = \Sigma x/n \tag{2.1}$$

TABLE 2.4  Tally of PPM Data in Ascending Order

| | | | | |
|---|---|---|---|---|
| 237 | 1 | | | |
| 238 | | | | |
| 239 | | | | |
| 240 | | | | |
| 241 | 1 | | | |
| 242 | | | | |
| 243 | 1 | 1 | | |
| 244 | | | | |
| 245 | | | | |
| 246 | | | | |
| 247 | 1 | | | |
| 248 | | | | |
| 249 | | | | |
| 250 | 1 | | | |
| 251 | | | | |
| 252 | 1 | | | |
| 253 | 1 | | | |
| 254 | | | | |
| 255 | 1 | 1 | 1 | 1 |
| 256 | 1 | | | |
| 257 | 1 | | | |
| 258 | 1 | | | |
| 259 | | | | |
| 260 | 1 | | | |
| 261 | | | | |
| 262 | 1 | | | |
| 263 | 1 | | | |
| 264 | 1 | | | |
| 265 | | | | |
| 266 | 1 | | | |
| 267 | 1 | | | |
| 268 | 1 | | | |
| 269 | 1 | | | |
| 270 | | | | |
| 271 | 1 | | | |
| 272 | | | | |
| 273 | 1 | | | |
| 274 | 1 | | | |
| 275 | 1 | | | |
| 276 | | | | |
| 277 | | | | |
| 278 | 1 | 1 | | |
| 279 | 1 | | | |

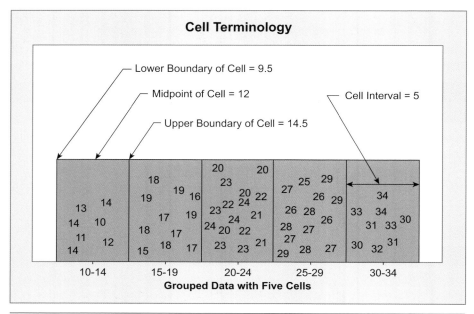

FIGURE 2.2    Cell terminology

so:

$$\bar{x} = (40 + 41 + 42 + 43 + 44 + 45) / 6 = 42.5$$

The midpoint for this cell is 42.5. In statistics, the average is known as the *mean*. Equation 2.1 will henceforth be referred to as the *mean equation*.

If possible, the number of cells should be odd and, as seen by the example above, an odd number within the cell helps as well. Regarding the number of cells, an odd number helps determine the middle cell needed for several statistical calculations. In determining the number of cells, the first consideration is how many will better support the distribution. If there are too few cells, the distribution will not be displayed adequately. If there are too many cells, the purpose of grouping is lost. The shape of the distribution will always be affected both by grouping and by the number of cells. There is no clear way to determine the best grouping. In the end, this decision rests with statistician as how to best represent the distribution.

The most common method of determining the number of cells is using what is called the *range* (highest reading minus lowest reading). First add one unit to the range; then divide by a desired number of cells in each interval. Try another desired number. Compare results and see which one best serves your purpose.

As an example, let's return to the data regarding the ppm of chlorine in the pond water. Table 2.5 shows there are now 120 raw data collected from the manufacturing

TABLE 2.5 PPM of Chloride in Pond Water (n = 120)

| | | | | | | | | | |
|---|---|---|---|---|---|---|---|---|---|
| **234** | 243 | 250 | 254 | 259 | 263 | 265 | 268 | 271 | 279 |
| 235 | 244 | 250 | 254 | 259 | 263 | 266 | 269 | 271 | 279 |
| 236 | 245 | 251 | 255 | 259 | 263 | 266 | 269 | 272 | 280 |
| 237 | 245 | 251 | 256 | 260 | 263 | 266 | 270 | 272 | 280 |
| 238 | 245 | 252 | 256 | 260 | 264 | 267 | 270 | 272 | 281 |
| 239 | 246 | 252 | 256 | 260 | 264 | 267 | 270 | 273 | 284 |
| 240 | 247 | 252 | 256 | 260 | 264 | 267 | 270 | 273 | 285 |
| 240 | 247 | 252 | 257 | 260 | 264 | 267 | 270 | 274 | 286 |
| 240 | 247 | 253 | 258 | 260 | 264 | 267 | 270 | 274 | 286 |
| 240 | 248 | 253 | 258 | 261 | 264 | 268 | 270 | 275 | 287 |
| 240 | 248 | 253 | 258 | 261 | 265 | 268 | 270 | 275 | 287 |
| 241 | 250 | 253 | 259 | 262 | 265 | 268 | 271 | 278 | **287** |

plant monitoring and arranged in ascending order. The highest and lowest readings are indicated in bold. The first step is to determine the number of cells we will use to group these data. The range is 53; we'll add one and work with 54. If we want 7 cells, we divide 54 by 7, giving us 7.7143, which would be rounded to 8. Thus, seven cells would deliver an interval of 8. In turn, 9 cells would deliver an interval of 6 whereas 11 cells would deliver an interval of 4.909, rounded to 5.

There are also some analytic methods to choose from, including Sturgis' rule, Rice's rule, and Scott's rule. The most common of these is Sturgis' rule, shown in equation 2.2:

$$k = 1 + 3.322\log_{10}n \tag{2.2}$$

This equation delivers an ideal number of cells (k) based on the number of readings in the sample (n). For this example with n = 120, Sturgis' rule delivers an answer of 7.907. This amount is rounded to 8 with the next odd number (of cells) being 9. Rounding up to 11 cells is also acceptable. Notice that the Sturgis rule gave results similar to the less analytic method mentioned above. Figure 2.3 shows the results of grouping these data with 9 cells and Figure 2.4 shows grouping these data with 11 cells.

Overall, these two graphs show little difference. The graph with 11 cells shows a slightly better accuracy in the distribution. The difference in statistical calculations between the two is negligible. The computed mean for the data using 9 cells is 261.2 whereas for 11 cells it is 261.125. Both round to 261, which then equals the ungrouped mean. If the distribution deviates from normal, these graphs would show a greater deviation in the distribution. Increasing the number of cells would help eliminate this deviation, showing a graph closer to reality.

Computing statistics using grouped data requires slightly modified equations from those used with ungrouped data. Chapter Four provides equations for both ungrouped and grouped conditions regarding those statistics.

Figures 2.3 and 2.4 also show good examples of histograms. Histograms are frequency charts showing the upper (and hence lower) boundaries of selected cells on the $x$ axis. The $y$ axis is the frequency. Also note that the $y$ axis is reduced as the number of cells increase.

Other similar frequency charts include a cumulative frequency chart, a relative frequency chart, and a combination of the two — a relative cumulative frequency chart. The relativity refers to the proportion of the data included in the cells to the total number of data; the proportions are typically stated in percentage in the $y$-axis. A cumulative histogram does not show the bell shape of the distribution. Instead, it shows the accumulation along each progressive cell until the total is represented. Figure 2.5 shows a relative cumulative frequency histogram of the same data. The $y$-axis is the relative scale and the numbers above the cell bars indicate the number of actual cumulative data (note the last cell indicates n = 120).

Other graphic representations of distributions include stem and leaf charts, box and whisker plots, and scatter plots. Examples of these and others follow.

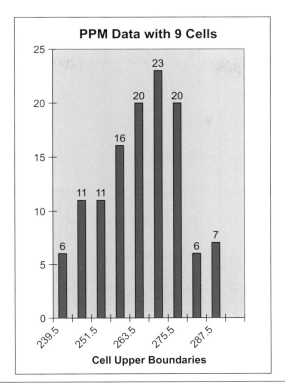

FIGURE 2.3  Grouping PPM data with 9 cells

## Stem and Leaf Plot

The stem and leaf plot is a quick method of organizing data into groups for graphic representation. This method is less analytic than formal grouping as shown above. As such, it does not allow accurate calculation of statistics. The main purpose of the plot is to show the shape of the distribution. Figure 2.6 shows a stem and leaf plot for the 120 PPM data in the example above where the data ranges from 234 to 287 (see Appendix E). The stem breaks each datum into a category based on its individual number. In this case, the first stem is 23; there are a few data in the 230s category. The remaining stems are 24, 25, 26, 27, and 28. The leaf refers to the remainder of the datum. For example, in the first stem category of 23, data in that category include 234, 235, 236, 237, 238, and 239. In the case of 234, 23 is the stem and 4 is the leaf. In the remaining categories, each datum is individually listed. For example, in the second stem category of 24, the number 240 is listed 5 times.

Note the similarity of shape to the histograms in Figures 2.3 and 2.4. As before, the number of stems determines the quality of the graph. If there are too few stems, the distribution will seem to be lumped together. If there are too many stems, the distribution will be spread out and not show the shape of the distribution.

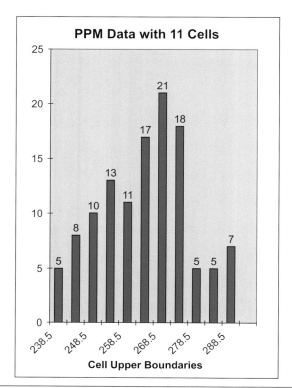

FIGURE 2.4  Grouping PPM data with 11 cells

To increase the graphic quality of some distributions, the stem can be split into different segments. This is called a split stem and leaf plot. One common technique is to split the stem into upper and lower segments. For example, if the stem labeled 24 is split, 24L will include 240 through 244, and 24U will include 245 through 249. Another technique is to split the stem into further segments such as "z" for zeros and ones, "t" for twos and threes, "f" for fours and fives, "s" for sixes and sevens, and "e" for eights and nines (Figure 2.7). These particular segment designations mean something in English, but be cautioned — they may be different in another language.

An ordered stem and leaf plot adds a third column to the plot to show how many data are below the median and how many data are above the median per segment. The *median* is defined as the middle datum. In the case of the PPM data there are 120 data. Because 120 is an even number, the midpoint between the 60th and the 61st data represents the median as 262.5. The new column shows the number of data in the stem containing the median. Each stem below the median shows how many data are in the stem and all the stems below it. Each stem above the median shows how many data are

Sampling and Organization of Data

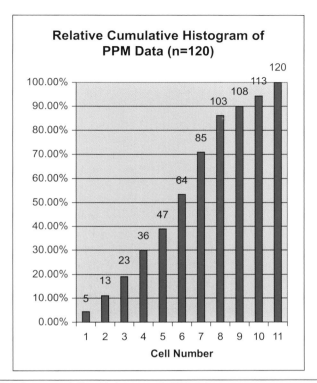

FIGURE 2.5  Relative cumulative histogram of PPM data

| Stem | Leaf |
|---|---|
| 23 | 4 5 6 7 8 9 |
| 24 | 0 0 0 0 0 1 3 4 5 5 5 6 7 7 7 8 8 |
| 25 | 0 0 0 1 1 2 2 2 2 3 3 3 3 4 4 5 6 6 6 6 7 8 8 8 9 9 9 9 |
| 26 | 0 0 0 0 0 0 1 1 2 3 3 3 3 4 4 4 4 4 4 5 5 5 6 6 6 7 7 7 7 7 8 8 8 8 9 9 |
| 27 | 0 0 0 0 0 0 0 0 1 1 1 2 2 2 3 3 4 4 5 5 8 9 9 |
| 28 | 0 0 1 4 5 6 6 7 7 7 |

FIGURE 2.6  Stem and leaf plot of PPM data

| Stem | Leaf |
|---|---|
| 23 z | |
| 23 t | |
| 23 f | 4 5 |
| 23 s | 6 7 |
| 23 e | 8 9 |
| 24 z | 0 0 0 0 0 1 |
| 24 t | 3 |
| 24 f | 4 5 5 5 |
| 24 s | 6 7 7 7 |
| 24 e | 8 8 8 9 9 9 9 |

FIGURE 2.7  Partial display of split stem and leaf plot of PPM sata

in that stem and all stems above. For example, in the 28 stem, the number 33 in the left column means there are 33 data in the 28 and 29 stems combined. Figure 2.8 shows an ordered stem and leaf plot using the PPM data.

Stem and leaf plots are easy to organize. Unlike histograms, where grouping or classification into cells is typical, stem and leaf plots can be made quickly; they give a similar graphic representation of the data. However, there may be fewer choices to represent a distribution graphically with a stem and leaf plot than with a grouped data histogram. Another advantage of stem and leaf plots (in particular the ordered stem and leaf plot) is the ability to see the amount of data in different areas of the distribution. This allows statisticians to describe the distribution and how the data are behaving in relation to variables.

A simple example of this is identifying quartiles and the inter-quartile range. A quartile is a quick measure of dispersion (see Chapter Four) where the distribution is divided into four equal segments. The number of data in each segment can show

| | Stem | Leaf |
|---|---|---|
| 6 | 23 | 4 5 6 7 8 9 |
| 23 | 24 | 0 0 0 0 0 1 3 4 5 5 5 6 7 7 7 8 8 |
| 61 | 25 | 0 0 0 1 1 2 2 2 2 3 3 3 3 4 4 5 6 6 6 6 7 8 8 8 9 9 9 9 |
| (median stem) 36 | 26 | 0 0 0 0 0 0 1 1 2 3 3 3 3 4 4 4 4 4 4 5 5 5 6 6 6 7 7 7 7 7 8 8 8 8 9 9 |
| 33 | 27 | 0 0 0 0 0 0 0 0 1 1 1 2 2 2 3 3 4 4 5 5 8 9 9 |
| 10 | 28 | 0 0 1 4 5 6 6 7 7 7 |

FIGURE 2.8  Ordered stem and leaf plot of PPM data

if the distribution is flat or steep. It can show to what extent the data group around the middle of the distribution. If the data are random and unbiased, there should be more data in the two middle quartiles (the inter-quartile range) and fewer in the end quartiles.

The disadvantage is having no clear numbers such as midpoints, cell intervals, and upper and lower boundaries with which to calculate the statistics of the distribution.

## Box and Whisker Plots

Box and whisker plots expand on the idea of quartiles. The box represents the inter-quartile range and the whiskers represent the end quartiles. Figure 2.9 shows a box and whisker plot for the pond water study with PPM data.

The plot centers around the median of the distribution. As noted above, the median is 262.5. This point becomes the middle line of the inter-quartile range (the box). The upper limit of the box (or third quartile) or the lower limit of the upper whisker is the median of the data above the middle line. Sometimes the upper and lower lines representing the inter-quartile range are called hinges. There are 60 data above the distribution median. Because 60 is an even number, the median of the upper half is

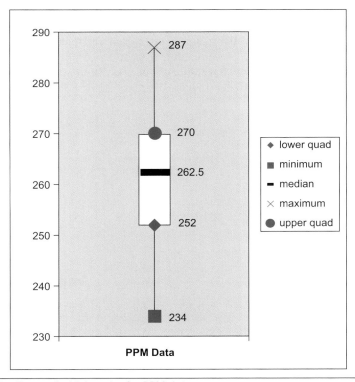

FIGURE 2.9  Box and whisker plot for PPM data

between the 30th and 31st numbers. In this example, these numbers are the same, and the upper limit of the box is set at 270. The upper whisker terminates at the highest datum of 287.

Similarly, the lower portion of the plot is calculated with the median of the lower half of the distribution. The lower limit of the box (or second quartile) is set at 252 and the lower whisker terminates at the lowest datum of 234.

Variations to the box and whisker plot include placing fences to indicate suspect outliers in the distribution. An outlier is a datum extending completely outside of what is considered probable in the distribution. It can be caused by a mistake in data recording, or some other variable present only to that datum. Outliers don't represent the distribution; many times, they are eliminated from the distribution.

The inner fence is located 1.5 times the inter-quartile range away from the hinges of the box. The outer fence is located 3 times the inter-quartile range away from the hinges of the box. Those data falling between the fences are considered possible or mild outliers. Those data falling outside of the outer fence are considered extreme outliers. Calculating fences for Pond 3 (Figure 2.10) shows neither of the high or low values are considered outliers. The inter-quartile range is 30.5. Therefore, the inner fence on the high side of the distribution is

$$(30.5 \times 1.5) + 245 = 290.75$$

and for the outer fence

$$(30.5 \times 3) + 245 = 336.5$$

The inner fence for the lower side of the distribution is

$$214.5 - (30.5 \times 1.5) = 168.75$$

and for the outer fence

$$214.5 - (30.5 \times 3) = 123$$

One clear advantage to the box and whisker chart is the ease of comparing multiple distributions. Suppose the PPM data used above represent data collected from only one of three ponds surrounding the manufacturing plant. If data from each of the ponds are plotted side by side, any similarities or differences will be graphically apparent, perhaps revealing an important aspect of the study (Figure 2.10). Pond 3 presents some statistical differences from the other two ponds. Its plot shows the distribution skewed (lopsided) to the low values. In this case, you might assume the lower PPM readings are desirable. So if the maximum PPM reading (280) is below the Environmental Protection Agency's limit, the manufacturing plant should duplicate the treatment of Pond 3 over the other two ponds.

## Scatter Plots

Scatter plots show a special relationship between the independent axis ($x$ axis) and the dependent axis ($y$ axis). Data are scattered across the plot and can reveal patterns or randomness of the distribution. They represent a change in the $y$ axis as related to

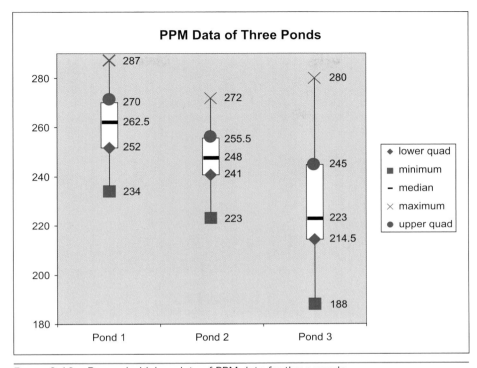

FIGURE 2.10  Box and whisker plots of PPM data for three ponds

a change in the *x* axis. To illustrate, consider a reaming operation in a manufacturing process. A high quality reaming tool result in longer use with higher-quality reamed surfaces. If use time is plotted as the independent axis (*x* axis) and quality of surface (measured in surface roughness) is plotted as the dependent axis (*y* axis), a high quality tool will show a random scatter. If a low-quality tool is used, the scatter will develop a pattern showing a dulling of the reaming blades.

Figure 2.11 shows the high-quality scenario depicting a random scatter of average roughness ($R_a$) data in microns measured 10 times every hour for 10 hours. The plot shows a random scatter over the time period, meaning the ream tool did not dull throughout the operation. Certain variables are kept constant such as use of same ream tool, type of material, piece dimensions, and treatment of material and hardness, as well as the feed and speed rate of the ream operation; there is also a consistency in data collection using the surface roughness gauge. Had the ream tool been of lesser quality, the scatter of the data would have shown a pattern that deteriorated over time, meaning the ream tool began to dull the more it was used.

Figure 2.12 shows how the scatter plot appears under a low-quality scenario. The pattern shown is called a positive curvilinear trend. However, being positive in this

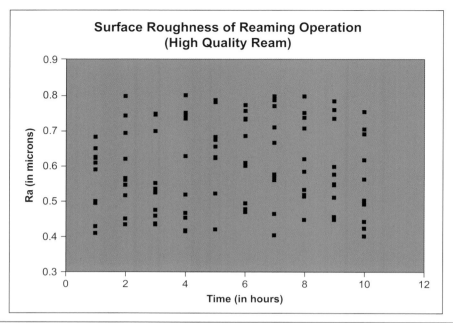

FIGURE 2.11   Scatter plot of reaming operation surface roughness — high quality ream

case does not denote a positive condition. It only means the curve is increasing to the positive side of the *y* axis. An increase in *y* indicates a greater roughness on the surface caused by a dull ream, which is undesirable. Other patterns to look for in scatter plots are linear (both negative and positive), negative curvilinear, and a variety of trends and cycles. These patterns will be covered briefly in a discussion of regression in Chapter Ten.

## Pareto Chart

In the late 19th century, Vilfredo Pareto observed, to the irritation of the Italian elite, 80% of wealth was held by 20% of the population. Later this became known in economics as the 80-20 rule. Not all scenarios fit neatly into this rule. However, many problems when grouped together tend to show relationships of the vital few to the trivial many. This phenomenon holds true within industrial systems as well. In labor management, there is likely a single greatest contributor to absenteeism. In supply management, chances are that a majority of purchasing or inventory is linked to just a few resources. In operations, the majority of quality issues may be caused by one or two problems with one machine or a single process. The Pareto chart graphically shows where the vital few problems exist, allowing managers and statisticians to concentrate

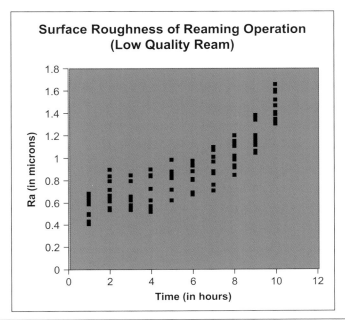

FIGURE 2.12   Scatter plot of reaming operation surface roughness – low quality ream

on those few. Focusing first on the vital few increases efficiency to a greater extent and faster than concentrating on the trivial many.

Consider a manufacturing process using a stamping machine. Quality problems with the product linked to this machine can be attributed to several issues. They include the machine's hydraulic system, the machine's pneumatic system, faulty gauges, sticky electrical control contacts, use of incorrect material, lack of operator training, and worn dies. After collecting adequate data on causes of machine breakdowns, these data can be charted to see the vital few causes opposed to the trivial many.

Figure 2.13 shows the Pareto chart for the study. Note that worn dies account for the majority of problems associated with this process, with 21 incidents over a 100-day period or 51% of the total number of incidents. Together with the machine's hydraulic system, operations can correct up to 71% of the problems with this process by concentrating on these two items. The other items are indeed problems that must be addressed; however, they are less critical than worn dies and hydraulics. Once the targeted problems are analyzed and corrected, a new Pareto chart can be constructed and the process continues. Each phase increases the efficiency of the operations process, creating over time an ongoing improvement in quality of the process and consequently the product.

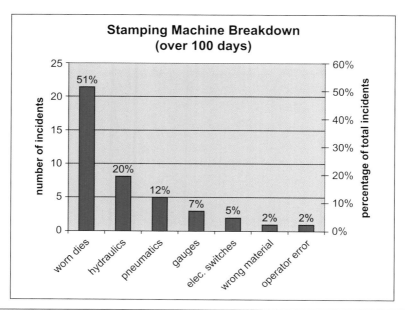

FIGURE 2.13   Pareto chart of stamping process

In actuality, the Pareto chart is nothing more than a bar graph. However, what the chart presents in the relation to grouped items defines it as a Pareto.

## Line Chart and Curves

Each of the charts or plots mentioned above can also be represented by a simple line. The scatter plots can be rendered into a single line with regression (see Chapter 10); the line represents a slope and perhaps a trend or reoccurring cycle. The histograms can also be placed into a line chart. Placing the PPM data for the chemical contamination of pond water study into a line chart creates what is called a *distribution curve* — meaning the production of a specific shape. The curve depicted in Figure 2.14 is a smoothing of the edges of the original histogram; it allows statisticians to see a clearer shape of the distribution. This distribution is not perfect. Rarely in practice does a study present a perfect distribution. The shape of the curve does show some semblance to what is called a *normal distribution* and can in fact be compared to a normal distribution for inferential purposes. A normal distribution (Figure 2.15) holds certain properties regarding the curve's height and width, a topic covered in more detail in Chapter Four.

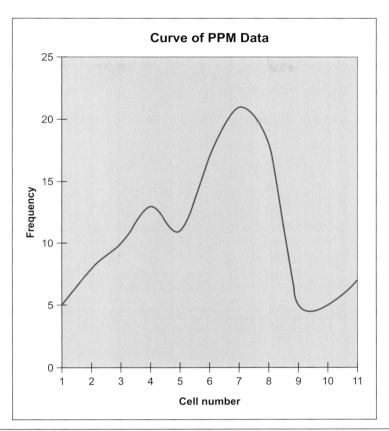

FIGURE 2.14    PPM data curve

As with curves, line charts give a clearer picture of the trend of the data. If the surface roughness data is fitted into a line chart, the high quality reaming scenario (Figure 2.11) will be a straight line. The low quality scenario produces a upward curve, as seen in Figure 2.16.

Other charts and graphs include variations of the bar charts use for histograms, pie charts, variations of the scatter plots, and a host of imaginative charts and graphs found in practice to get a graphic representation of data and what they stand for.

Graphs, charts, and any other form of graphic representation of data or a distribution is only half the story. These graphic forms of statistics are only viable with analytic measures. Used in conjunction, these charts and graphs can better describe distributions and can aid in making important decisions in the industrial setting.

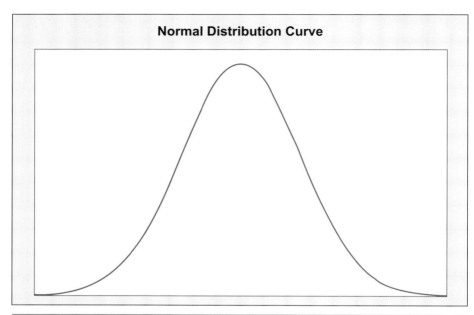

FIGURE 2.15  Normal distribution curve

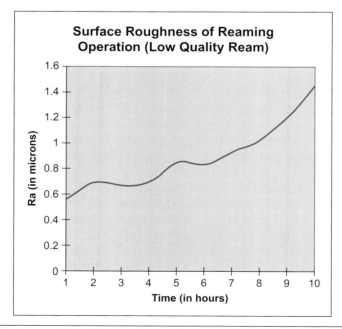

FIGURE 2.16  Low quality surface roughness line chart

## Chapter Two Discussion

1. How are statistics used in industry?
2. What is the difference between a population and a sample?
3. What is raw data?
4. How is sample size determined when collecting data for a statistical study?
5. What is meant by random? Give an example of random data as opposed to non-random data.
6. What is bias? How can bias affect a statistical study?
7. What is meant by controlling the variables?
8. For what purpose is a Pareto Chart used?
9. What are the advantages and disadvantages of using box and whisker charts or scatter plots?

## Chapter Two Problems

1. Using the following data: 13, 17, 10, 19, 13, 18, 11, 13
   a. Tabulate into an ascending array and a descending array.
   b. Using the ascending array, tally the data.
   c. Draw a histogram of the data. Try using Minitab® or Excel® to create a histogram.
2. Why do statisticians group data into cells? Give some examples. Make up your own data (about 100) and group them into cells. Repeat the exercise with a different cell interval.
3. Enter the data you generated in question 2 into Minitab® or Excel®. Create a histogram to show these differing intervals. Also hand draw the histogram.
4. Make a relative frequency histogram of the data used in question 2.
5. Using the provided data set for this problem (see Appendix E), create in Excel® and/or Minitab® each of the following:
   a. Tabulation
   b. Tally
   c. Histogram
   d. Grouped histogram using 5, 7, and 9 as intervals
6. Using the data from problem 5, create a stem and leaf plot and an ordered stem and leaf plot in Minitab®. Also create the plots by hand.
7. Using both Excel® and Minitab®, create a box and whiskers plot of the data given in problem 5. Practice creating them by hand.
8. Create a curve of the data in problem 5 using Excel®, and also by hand.

# THREE

## Basic Statistics for Discrete Distributions

Without realizing it, people use the language of probability every day. Anytime someone uses words such as *likely*, *likelihood*, *probably*, *chances are*, *maybe*, or any other word denoting a degree of occurrence, they are speaking the language of probability. When they use the language of probability, they are using the language of statistics because all statistics are a form of probabilities. *Probability* relates to occurrences. A probability of 0.00 indicates impossibility whereas a probability of 1.00 indicates a certainty. Any value in between — greater than 0.00 but less than 1.00 — indicates a possibility. The higher the value is, the greater the likelihood (or chance) that an event will occur.

Because the language of statistics is probability, a basic understanding of statistics starts with an understanding of probability. Probabilities are generally derived by three methods: classical, relative, and subjective. Variations in probability develop based on how events and outcomes are counted. *Outcomes* are the result of a trial, such as throwing a die. A successful outcome, such as a six from a thrown die, is considered an event. *Events* can be a single successful outcome or may represent a series of multiple outcomes. Understanding how probabilities develop into distributions is critical to understating estimations, data sets, and further use of statistics.

## Classical Probability

Classical probability is the most accurate of all probabilities. A classical probability can be calculated specifically without influence from past experience or judgment. In other words, an event or specific outcome is predetermined by using the probability equation shown below.

$$P(E) = \text{successful outcome} / \text{total possible outcomes} \qquad (3.1)$$

where P is the typical statistical notation for probability and E denotes event.

The most common examples of classical probability are flipping a coin, drawing cards, or throwing a die. In each of these cases, all variables must be equally controlled and non-biased. For example, if a die is loaded on one side, tossing or tumbling the die adequately will result in the heavy side ending down on the table a greater number of times, if not always. The same is true with a coin. A well-balanced coin, or a fair coin, has an equal probability of turning up heads or tails. Because there is only one of two outcomes, the equation uses 1 as the successful outcome or event (turning up heads in this case), and 2 as the total number of possible outcomes (turning up heads or tails). Using equation 3.1, the probability of tossing a fair coin with an outcome of heads is:

$$P(heads) = 1/2 = 0.5$$

Likewise, with the toss of a die, the equation uses 1 event out of 6 possible outcomes. Each toss of a fair die returning a six has a probability of:

$$P(six) = 1/6 = 0.1667$$

Moving on to a more complicated example, consider a deck of playing cards. There are 52 cards in a deck consisting of 13 spades, 13 hearts, 13 diamonds, and 13 clubs. The probability of drawing a single card in a fair shuffled deck and getting a spade is:

$$P(spade) = 13/52 = 1/4 = 0.25$$

Thus the probability is one event out of 4 possible outcomes; and this probability reads as 0.25. Drawing a specific card such as the ace of spades is:

$$P(ace\ of\ spades) = 1/52 = 0.01923$$

Putting these calculated results in perspective regarding the language of probability is simple. Drawing an ace of spades out of a fair shuffled deck has a probability of 0.01923. We can specify the probability in terms of chance, for example, saying a probability of 0.01923 has just under a 2% chance, or more specifically a 1.923% chance. Similarly we can say the odds are 1 out of 52.

In industry, classical probability is limited in use because product is produced either in lots or continuously over period of time. The probability (of a defect, for example) must be calculated over a long period of time using several trials and averaging the events. The more trials in the experiment will bring relative probability closer to a classical probability.

## Relative Probability

A relative probability is the occurrence of an outcome or event over a period of time or given number of trials. If a fair coin is tossed once, the probability of heads is 0.5 and the probability of tails is 0.5. If the coin is tossed 10 times, we assume heads will turn up 5 times and tails will turn up 5 times. This is not necessarily so. Try this experiment and record the results. Suppose the coin is tossed 10 times resulting in 6 heads and 4 tails. This gives a relative probability of 0.6 of receiving heads. According to classical probability, this probability should be 0.5. Now, repeat the experiment 9 more times (toss the coin 90 more times) and compute the relative probability for each of the sets of 10. As more tosses are completed and probabilities calculated, the closer the relative probability is to the classical probability of 0.05. Theoretically, as the number of coin tosses approaches infinity, the relative probability comes closer to the classical probability. This sort of probability is relative probability, sometimes called experimental or historical probability. Even with the absence of a classical probability scenario, this method of calculating probability can reveal reliable probabilities if enough trials are performed.

Relative probability is used extensively in industry, where product is produced over a long period of time. Customers, such as merchandise companies or manufacturing plants buying product from a supplier, look at relative probability to further calculate the probability of ending up with product from a specific lot that does not conform to expectations. Manufacturers look at relative probability when compiling control charts for quality control through the manufacturing process. Both attribute and variable charts operate on the same principal (see Chapters Five and Six) where an average of the distribution is an established relative probability that product will migrate toward the center. Relative probability is also used in operations management for inventory control. Stocking levels and reorder amounts are justified based on the historical demand for the product qualified by a level of service to the customer. Each are typically determined using a relative frequency.

Suppose a manufacturing plant producing electric motors wishes to determine the failure rate of the winding in the stator section of the motor. This component is produced continuously during eight-hour shifts, 7 days a week, 365 days a year. The plant recently concluded an experiment that tested stators for continuity by applying a set amperage with a set voltage to the coil leads. During the experiment, which lasted two years, conforming product withstood the test for 10 seconds whereas nonconforming product failed in dielectric strength. Results were recorded as pass or fail for each stator leaving the process. Out of 1.5 million stators, 27,000 failed. Calculating the relative probability gives a failure rate of:

$$P(\text{stator failure}) = 27{,}000/1{,}500{,}000 = 0.018$$

or just under 2 %.

Now suppose a customer for the stators receives a shipment of 4 lots where a lot is 1728 units. How many nonconforming stators will the customer expect to find in each lot? Using the established relative probability of 0.018, each lot should have no more than

$$1728 \times 0.018 = 31.104 \text{ rotors}$$

For the entire shipment, the customer should expect

$$(1728 \times 4) \times .018 = 124.416 \text{ rotors}$$

Stators, of course, do not exist in fractions — it is impossible to have 0.104 or 0.416 of a stator. To make sense, these numbers must be rounded. However, if either of these numbers are rounded up, the amount of nonconforming product will exceed the relative probability of the study. Therefore, for this shipment, the customer expects 124 nonconforming stators. This example demonstrates an example of a discrete distribution in which the measurement of a unit is limited to a whole number; a topic which will be discussed further in this chapter.

The relative probability of stator failure can be used by the producer for evaluation purposes and quality improvement. Machine operators who produce 1000 stators where 30 are bad (3% fail rate) should improve their performance. Meanwhile, operators who produce 1000 stators where 10 are bad (1% fail rate) should be giving seminars on how they have improved the process.

## Subjective Probability

Subjective probability refers to a degree of occurrence based on anything from a gut feeling (or intuition) to an expert opinion. It has no mathematical formula; however, it can result in a number congruent with other probabilities. Remarkably, subjective probability is used quite extensively in industry. During evaluation activities, marketing ventures, purchasing agreements, and other situations, a subjective probability of success or failure can turn a decision one way or another. This process should not necessarily suggest a series of sub-standard approaches to making decisions; rather, it should point toward what is often a valuable qualification of the situation. One disturbing trend in business and industry today is that of relying too heavily on purely quantitative data. Subjective approaches complete a greater picture of what might be closer to reality. The ideal approach is a combination of all sources of probability.

Let's consider an example of horse racing to see how these types of probabilities can work together. What are the variables during a race that determine a winner? A racing brief from the track includes a listing of the horses and their conditions, the jockeys, the track, the weather, the competition, the length of the races, and other information including how other bettors feel the race will unfold. This short list includes relative, subjective, and classical probabilities. The horse alone is subject to several variables. How old is the horse? How long has the horse been racing? How many races did the horse already run that day (could suggest fatigue)? How does the horse perform on the condition of the track? Answers to many of these questions most likely result in a relative probability. Then there are some purely subjective considerations. What does the horse look like? Even horses where high relative probabilities have been calculated for some specific event may be having a bad day. The professional horse bettor has extensive knowledge of horses and can observe and detect those conditions — perhaps overriding that high relative probability.

All of these questions are considered and then quantified into yet another form of chance — that of odds generated through the pari-mutuel betting system. This system is how the bettors see the odds of a particular horse in a race and calculate how much money they may win (or lose). This system considers how many people are betting on which horse and establishing favorites and long shots on how the horses will perform. In addition to all of this information, there is the classical probability of how many horses are in the race. If there are 10 horses racing, only one can come in first. This gives a classical probability of .01. However, three horses will make money. One horse will win, another horse will place, and another will show. Choosing three horses, all to show, will give a 3/10 = 0.3 classical probability of making money. This classical probability of 0.3 combined with all the other relative and subjective probabilities mentioned above tends to increase the total probability to greater than 0.3. How much greater than 0.3 is a subjective guess.

In industry, sources of variation occur during the process at unpredicted times, leading to periodic use of subjective probability. The variables must constantly be monitored and decisions made in regard to the effect they have on production. Those individuals with extensive experience and knowledge within that industry will make or at least influence those decisions.

## Probability Variations

Each type of probability uses certain terms and rules consistently. An *experiment* generally involves a series of trials, but can also be just one. A *trial* is a single action to acquire a degree of occurrence or probability. An *outcome* is the result of a single trial. An *event* is the occurrence of a successful outcome; it can also represent outcomes from several trials or a single trial with multiple outcomes. Probabilities are always expressed as a number between 0.00 and 1.00. Occasionally, other terminology is used such as *chance* (typically denoted with a percent) and *odds* (typically denoted with a ratio). Because the probability of a certain outcome is P(x), the probability that P(x) will not occur is 1.00 minus P(x). The sum of all possible outcomes or events totals 1.00.

When calculating probabilities, there are certain variations given different situations. These variations have to do with multiple events through single or multiple trials as well as the relationship that outcomes and events have or do not have on one another. In addition, situations rise where more complicated means of counting the total number of possible outcomes is required to calculate probability. Consider the following examples.

### Mutually Exclusive (One Event)

When one event prevents another possible event from occurring, the two events are said to be *mutually exclusive*. Heads resulting from the toss of a coin prevents the event of tails. A die cannot show a 6 and any other number at the same time. The probability is calculated the same way as P(6) = 1/6 = 0.167, but it becomes a mutually exclusive probability.

Consider an electric motor plant. One operation in the process produces the laminated core of the rotor. In this process, stamping machines punch out several circular cross-sectional pieces for the core. These pieces are stacked together, skewed to the correct degree, and inserted into a casting process to form the rotor. There are three machines producing the laminates. Two machines are old and the laminates fall out the bottom of the stamping machine where they are gathered, stacked to a certain weight, and skewed by hand. The other machine uses an interlocking system where each laminate is locked to the other with the correct skew and stack weight automatically. These rotors are then cast and used in the assembly of the final motor.

The automatic machine (machine A) produces 500,000 laminated rotor cores per year; 6000 of those are non-conforming. Each manual machine (machines B and C) produces 125,000 rotor cores per year; 4375 from each machine are non-conforming. It is evident the plant would like to upgrade all laminate operations to the automated machine, but to keep up with capacity they must keep the old machines until upgrading is feasible. Customers have become aware of these differences and know that rotors from machine A are superior in quality.

What is the probability of a getting a rotor from either machine B or C? Machine B produces 125,000 out of 750,000 rotors. This gives P(machine B) = 125,000/750,000 = 0.167. Likewise, machine C produces 125,000 out of 750,000 giving P(machine C) = 0.167. The probability of getting a rotor from machine B or C is:

$$P(\text{machine B or C}) = P(\text{machine B}) + P(\text{machine C})$$

or:

$$P(\text{machine B or C}) = 0.167 + 0.167 = 0.334$$

Note the *or* term used in this model. This denotes the mutually exclusive condition. In the event a rotor from machine B is selected, it prevents a rotor from machine C being selected. This condition is sometimes referred to as the addition rule in mathematical probability.

## Not Mutually Exclusive (One Event)

Not mutually exclusive is the negation of mutually exclusive. The occurrence of one event does not make another event impossible. Drawing a spade or a king from a deck of cards is an example. The probability of drawing a spade is P(spade) = 13/52 = 2.6. Getting a king (of any suit) is P(king) = 4/52 = 0.077. However, one of those kings is a spade and was already accounted for in the P(spade) equation. Therefore, the probability of drawing either a spade or a king would be the addition of P(spade) and P(king) minus P(king of spades), or,

$$P(\text{spade, king, or both}) = (13/52 + 4/52) - 1/52 = (2.6 + 0.077) - 0.019 = 2.658$$

Note the use of the term *or* again. This case still uses the addition rule; however, there is a special condition. When non-mutually exclusive conditions apply, the probability of that which is common between the events must be subtracted, as in the case above with the king of spades.

Using the rotor example again, what is the probability of getting a rotor from machine B or C, or a non-conforming rotor? Non-conforming rotors can come from all three machines. However, the customer wants neither a non-conforming rotor from any machine nor rotors from machines B and C because of their perceived low quality. The probability of getting a rotor from machine B or C remains the same at 0.334. The probability of getting a non-conforming rotor is

$$P(\text{non-conforming}) = 14750/750000 = 0.0197$$

Of these, the probability of getting non-conforming rotors from machines B and C is

$$P(\text{non-conforming from B and C}) = 8750/750000 = 0.0117$$

The latter must be subtracted from the total probability to calculate the correct answer. So:

$$\begin{aligned}P(\text{machine B or C or non-conforming}) \\ = [P(\text{machine B}) + P(\text{machine C}) + P(\text{non-conforming})] \\ - P(\text{non-conforming from B and C})\end{aligned}$$

or

$$\begin{aligned}P(\text{machine B or C or non-conforming}) = (0.167 + 0.167 + 0.0197) - 0.0117 \\ = 0.342\end{aligned}$$

Because P(machine B) and P(machine C) include both conforming and non-conforming rotors, the probability of non-conforming rotors from those machines must be subtracted. This shows the non-mutually exclusive relationship between machines B, C, and non-conforming rotors. In other words, it is possible to have a rotor from machine B or C and that rotor be non-conforming.

Much of the use of mutually exclusive and non-mutually exclusive is dictated through terminology when defining the probability. This probability could just as easily be called a mutually exclusive situation and solved as:

$$\begin{aligned}P(\text{machine B or C or non-conforming from A}) \\ = P(\text{machine B}) + P(\text{machine C}) + P(\text{non-conforming from A})\end{aligned}$$

or:

$$P(\text{machine B or C or non-conforming from A}) = 0.167 + 0.167 + 0.008 = 0.342$$

where

$$P(\text{non-conforming from A}) = 6000/750000 = 0.008$$

## Independent versus Dependent

Independent events are outcomes that have no influence over subsequent outcomes. This is commonly called multiple trials without replacement. Consider the typical lottery system for those players that may be interested in knowing how to figure the probability of winning. Some games such as Pic 3 or Pic 4 are simple games where the player chooses a number between 0 and 9. If these numbers are selected, the player is a winner. There are 10 numbers between 0 and 9 giving a probability of $P(0 - 9) = 1/10 = 0.10$. This represents the first of however many trials is required for the game (3 in the case of Pic 3). After the first trial, the number is replaced back into the selection process ready for the second trial drawing. This means it is possible to get the same number twice, or even thrice.

Because the number is replaced, the first trial has no effect on the second trial or the third trial. So, the probability for the second trial remains the same as the first trial, P(0 – 9) = 1/10 = 0.10. And the third trail is the same. However, because the object of the game is to select the "lucky" numbers, the player must match those winning numbers selected by the lottery system. What is the probability the player will select the same number as the winning number? Again, P(0 – 9) = 0.10. Having matched the winning number on the first trial, what is the probability the player will match the winning number twice, or thrice (in the exact same order)? The probability is calculated as follows:

$$P(pic3) = P(0-9) \times P(0-9) \times P(0-9)$$
$$= 0.10 \times 0.10 \times 0.10$$
$$= 0.001$$

This means 1 in 1000 players are likely to win. It means a single player who plays 1000 times might win once. With payoffs less than $1000, it seems a player really must be lucky to get ahead. It is, after all, a probability — a relative probability. There are no guarantees the player would win after 1000 games, and conversely, the player may hit the jackpot the first time.

Some of the higher-prize games such as Power Ball require a slightly different approach. Choosing five numbers from 1 to 55 gives a probability during the first trial of

$$P(1-55) = 1/55 = 0.018$$

Because the number is not replaced (cannot be repeated), the probabilities of each subsequent trial are influenced by the preceding trials. This illustrates how each preceding trial can influence succeeding trails. In other words, it illustrates a dependent event. For example, the probability for the first five numbers are :

$$P(5 \text{ numbers}) = 1/55 \times 1/54 \times 1/53 \times 1/52 \times 1/51 = 0.0000000024$$

This is the probability of winning if the numbers must be in the exact same order. To calculate the probability where the winning numbers can be in any order, consider the following:

$$P(5 \text{ numbers any order}) = 5/55 \times 4/54 \ 3/53 \ 2/52 \times 1/51 = 0.00000029$$

By adding an independent Power Ball number between 1and 42, this value becomes:

$$P(5 \text{ numbers any order with 1 power ball}) = 5/55 \times 4/54 \times 3/53 \times 2/52 \times 1/51 \times 1/42$$
$$= 0.0000000068$$

To summarize, 1 out of 146,108,047 sets of numbers will likely win. It is a probability more useful to the Lottery Board than the player. Over a long period of time, the Lottery Board will know how much they will pay out. It does not mean that a player who plays 146,108,047 times is guaranteed to win. Note the descending denominators in both Power Ball examples for numbers in exact order and numbers in any order. This again shows the dependent nature of the probability and results from not replacing the number after the trial. For independent events, the denominator will remain steady.

## Counting Possible Outcomes and Events

As seen in the examples above, counting possible outcomes is quite easy in certain situations. The lottery example had 55 numbers from which to choose. The first trial offered 5 chances to match a number — hence, the diminishing numerators in subsequent trials. Occasionally, however, there are situations where determining the number of possible outcomes is more complicated. Knowing the number of all possible outcomes is required before probabilities can be calculated.

The simplest of these is multiplication. For example, a brewery needs to collect 15 samples from each of their 5 types of beer (for example, pilsner, ale, bitter, stout, and extra stout). How many samples will they have? Fifteen multiplied by five is 75. After sampling three types, they have $15 \times 2 = 30$ more to go. Other situations have a combination of possibilities; they require combination or permutation equations involving the use of factorials. (Factorials use the symbol ! to denote multiplication of the given number in a descending order, as shown by the following examples.)

### Combinations

*Combinations* determine the total number of possible groups, given how many individual objects are in the group and how many individual objects are in the sample from which the groups were taken. The general terminology for a combination is the number of objects taken so many at a time. Consider the word "ACE." With three objects (A, C, and E) selected two at a time, the combinations are AC, AE, and CE. All objects are represented only once, giving a total number of combinations of three. The equation is as follows:

$$C_{n,r} = n! / [r! (n-r)!] \qquad (3.2)$$

Where n = total number of objects and r = the number of objects within the group. For the "ACE" problem, this means:

$$C_{3,2} = 3! / [2! (3-2)!] = (3 \times 2 \times 1) / [(2 \times 1) \times 1] = 6/2 = 3$$

where the 3 in $C_{3,2}$ is A, C, and E, and the 2 in $C_{3,2}$ is selecting two letters at a time.

A more pertinent example for industry may be how many ice cream flavor combinations an ice cream manufacturer may devise by combining two flavors at a time from a total of twelve possible flavors (i.e. vanilla and cinnamon, nutmeg and ginger, vanilla and ginger, etc.). This calculation is:

$$\begin{aligned}C_{12,2} &= 12! / [2! (12-2)!] \\ &= (12 \times 11 \times 10 \times 9 \times 8 \times 7 \times 6 \times 5 \times 4 \times 3 \times 2 \times 1) / [(2 \times 1) \\ &\quad (10 \times 9 \times 8 \times 7 \times 6 \times 5 \times 4 \times 3 \times 2 \times 1)] \\ &= (12 \times 11)/2 = 66\end{aligned}$$

The ice cream manufacturer can determine the probability of one, or more, of these 66 flavors actually going to market. Suppose a panel of five tasters delivers a pass or fail decision on each flavor combination. For the flavor combination to go into production, the five panelists must unanimously agree. The pass or fail assessment denotes a 1/2 or 0.5 probability of a mutually exclusive situation. The probability of one flavor combination being selected is determined by the five panel members with no effect on the other flavor combinations; therefore, this probability is an independent scenario. Each flavor combination goes to each panel member with a 0.5 probability. As the flavor combination moves to each subsequent member of the panel that probability is reduced as follows:

$$P(\text{acceptance}) = 1/2 \times 1/2 \times 1/2 \times 1/2 \times 1/2 = 0.03125$$

There is a 0.03125 probability one flavor combination will make it through the entire panel and be introduced to the market. Each subsequent flavor combination will have the same probability as the first. Because there are 66 flavor combinations, the most likely number of selected flavor combinations is:

$$N(\text{selected flavors}) = 66 \times 0.03125 = 2.06 \text{ (rounded to 2)}$$

This number helps the company in areas of operations, marketing, and finance by allowing them to tool up for production, procure resources and supplies; allocate capital expenses; and formulate effective advertising for most likely two, but possibly three, new flavors of ice cream.

Suppose the CEO of the company wants to determine four selections by personally tasting the flavors, arguing this method would be just as effective as a panel. The probability a panel of five (let alone the market) will select the same four-flavor combinations that the CEO selects (a dependent event) is:

$$P(4) = 4/66 \times 3/65 \times 2/64 \times 1/63 = 0.00000137$$

With this low probability, we can conclude the best method of determining the flavor combinations introduced to the market come from a panel, or perhaps a market test, rather than one person such as the CEO.

Incidentally, the probability of all flavor combinations being selected by the panel is:

$$P(\text{all}) = 0.03125^{66} = 4.572 \times 10^{-100}$$

Although theoretically this is still possible, from a practical point of view it is very close to impossible and would be considered so in industry.

## Permutations

In combinations there is no assigned order to the objects as long as they are represented. Permutations find the number of possible combinations in every order possible. Instead of ice cream, imagine a blend of substances where the product is sensitive to the order in which the ingredients are added to the mix. For example, a metal alloy or a chemical compound may involve several pure materials. Each addition of a new material may change the physical, thermal, or chemical properties of that mix, perhaps inhibiting or enhancing the ability of the mix to assimilate subsequent additions. In other words, if item A is inserted prior to item B, or B inserted prior to C, the product will be different. Combinations of ACE, selected 2 at a time where order does not matter, include AC, AE, and CE. However, when order does matter, given the permutation equation,

$$P_{n,r} = n! / (n - r)! \qquad (3.3)$$

permutations of ABC (3 objects selected 2 at a time) include a total of:

$$P_{3,2} = 3! / (3 - 2)! = (3 \times 2 \times 1) / 1! = 6$$

Knowing there are six total permutations, we can identify and account for each of them. In the case of ACE, the permutations are AC, AE, CE, CA, EA, and EC for a total of six.

## Discrete Distributions

As discussed toward the end of Chapter One, variable measures are different than discrete measures. A variable measure, one with infinite decimal places of accuracy, fits into a continuous type distribution. These include measures such as weight, length, or volts. Can a volt meter give a reading of 12.56749238 volts? Yes it can, indicating the measure is continuous. Discrete measures, on the other hand, are items such as cars, planes, people, or whole units with countable numbers (1, 2, 3, 4…). From a practical standpoint, it is not possible to have one-half of a person or a car.

A *distribution* is a group of collected data arranged in order to reveal characteristics about what is being studied (Chapter Two). One of these basic characteristics is determined by the type of measurement — whether it is variable or discrete. For variable measures, collected samples generate continuous distributions (see Chapter Four). For discrete measures, collected samples generate discrete distributions. Suppose a random sample taken of eighteen people is compared to a greater population. Out of the eighteen people, half are men (the sample has a total of nine men and nine women). In the population, 25% have dark hair. But 25% of the sample would be 4.5 people, which is not possible. More important, if the sample has 4 people with dark hair, one might assume incorrectly there are N(people with dark hair) = 4/18 = 22.222% of the population with dark hair.

In general, distributions are sets of measures to show the probability of a certain criterion. This is true for both continuous and discrete distributions. Distributions can be generated by a sample, or they can represent a known population. There are also standardized distributions used to describe a population. Sample distributions are frequently compared to standardized distributions to describe and infer characteristics about a particular population. The distinguishing characteristic of a discrete distribution is the incremental step caused by the discrete measure (people, cars, planes, etc.). The resulting curve generated by a sample may resemble Figure 3.1, which illustrates the ages of children in a group. This group has 9 four-year-olds, 5 three-year-olds, 2 two-year-olds, and so on. The step between those values indicates, for example, there cannot be 1.5 children at the age of two.

**54**  Chapter Three

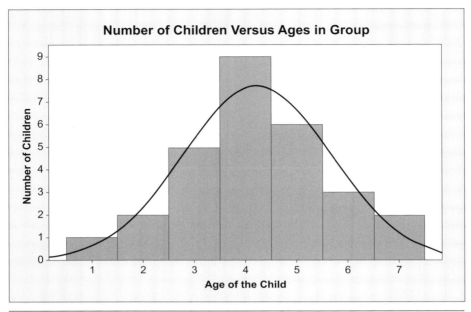

FIGURE 3.1  Characteristics of a discrete distribution

Discrete distributions are generated based on how big the sample is and in consideration of time constraints. The three most commonly used discrete distributions are the hypergeometric distribution, the binomial distribution, and the Poisson distribution. Prior to using any distribution in statistics, you should be aware of the *assumptions* for each distribution. In industry, assumptions serve as conditions. In other words, these certain conditions must be met before the distribution can be considered valid to the study. Accordingly, each of these distributions assumes a binary outcome during the trial where the trial probability is 0.5 (for example, 0 or 1, go or no-go, heads or tails); they also assume trials are independent. As covered earlier in this chapter, independence refers to each individual trial having no influence on subsequent trials. For example, the probability of heads or tails (0.5) remains the same for each trial — there is replacement of objects during the trial. Although mass-produced items are not actually replaced back into the production line to be measured again, the infinite nature of constant production accounts for this same assumption. It is sometimes referred to in industry as an infinite supply.

## The Hypergeometric Distribution

In situations where few or a finite number of objects are available for a sample, what is called in statistics as the *hypergeometric* distribution distinguishes a curve of probabilities. The hypergeometric equation finds the probability using the following combinatorial equation:

$$P(\text{\# non-conforming}) = C_{D, k} \, C_{N-D, n-k} \, / \, C_{N-n} \tag{3.4}$$

where

$D$ = total number of nonconforming objects in the population
$k$ = nonconforming objects drawn during the trial
$N$ = total number of objects in the population
$n$ = number of objects drawn during the trial (sample size)

Consider a coin stamping process where a muling error occurs (using two different dies for the same coin, for example, the head of a Washington quarter and the tail of a Sacajawea dollar). The machine operator notices the error and immediately halts operations. The clean-up crew determines 10 coins were stamped incorrectly but still fell into the bin currently holding 50 coins (including the non-conforming coins). Assuming the bin is well mixed and a handful of 5 coins are randomly selected from the bin, what is the resulting probability distribution covering each of the 6 possibilities from selecting 0 non-conforming coins to 5 non-conforming coins? Given the hypergeometric equation, the probabilities of finding respective non-conforming coins in the sample is

$P(0) = C_{10, 0} \, C_{50-10, 5-0} \, / \, C_{50, 5} = 0.310562782$
$P(1) = C_{10, 1} \, C_{50-10, 5-1} \, / \, C_{50, 5} = 0.431337197$
$P(2) = C_{10, 2} \, C_{50-10, 5-2} \, / \, C_{50, 5} = 0.209839718$
$P(3) = C_{10, 3} \, C_{50-10, 5-3} \, / \, C_{50, 5} = 0.044176783$
$P(4) = C_{10, 4} \, C_{50-10, 5-4} \, / \, C_{50, 5} = 0.003964583$
$P(5) = C_{10, 5} \, C_{50-10, 5-5} \, / \, C_{50, 5} = 0.000118937$

These calculations represent one trial where five coins are selected without replacement (Trial 1) and the probability of any individual coin being non-conforming is 0.5. The sum of the probabilities equals 1.00. Graphically, Figure 3.2 shows the distribution of this trial leaning toward the left. The highest point on that graph is where the expected value (mean, or majority of the data) is 1 coin. Conducting yet another trial is considered a dependent event because no coins are replaced in the bin after the first trial. The parameters change depending on the results of the first trial. Suppose during Trial 1, 5 non-conforming coins are found with a probability of 0.000118937. What is the probability of finding all 10 non-conforming coins within two handfuls? The total in the bin is now down to 45. The total remaining non-conforming coins is now 5. To find the probability of selecting 5 non-conforming coins in the second trial (Trial 2), the equation is modified as follows:

$$P(5) = C_{5, 5} \, C_{45-5, 5-5} \, / \, C_{45, 5} = 8.18492\text{E-}07$$

Multiplying the probabilities gives:

P(all errors in 2 trials) = $0.000118937 \times 8.18492\text{E-}07 = 9.7349\text{E-}11$

Clearly this scenario is not the most likely one. The most likely scenario is finding 1 non-conforming coin (the expected value) with a 0.431337197 probability during

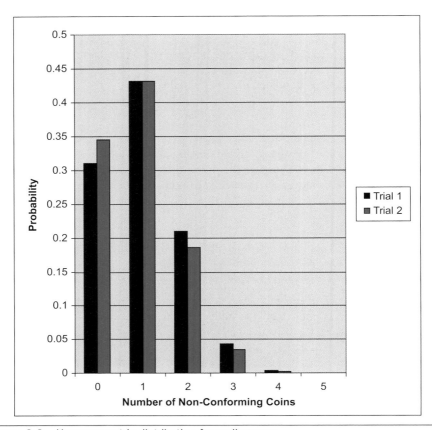

FIGURE 3.2  Hypergeometric distribution for muling error

the first trial and the same probability during the second trial. The shape of the distribution changes slightly, however, with the second trial. The leaning to the left is even more pronounced but the expected value is still 1 coin (Figure 3.2, Trial 2). This leaning toward the left (skew) will continue until all non-conforming coins are retrieved.

## The Binomial Distribution

The binomial distribution is similar to the hypergeometric distribution except the population is infinite. There is an infinite supply of objects for which the probability of non-conforming product is known through historic records of the process (for example: a relative probability established over several years). The other assumptions remain the same. The criterion for each object is conforming or non-conforming, resulting in a probability of 0.5. The sequence of selection during the trial is independent because of the infinite supply. Therefore, replacing objects or not replacing objects once they are selected is normally not an issue in practical application. Because the total to choose from is infinite, larger sample sizes are common; the resulting curve of the distribution becomes smoother. The fact still remains, however, that the distribution is discrete and the objects are counted in whole numbers. Calculating the probability for a single binomial is accomplished as follows:

$$P(\# \text{ non-conforming}) = n! / k!(n-k)! \, p^k (1-p)^{n-k} \qquad (3.5)$$

where:

$p$ = known probability of non-conforming objects in the population
$k$ = nonconforming objects drawn during the trial
$n$ = number of object drawn during the trial (sample size)

Note the total number of objects in the population (N) is not identified because of the infinite supply. The total number of non-conforming objects in the population (D in the hypergeometric distribution) is now replaced with p; a known value derived by a relative probability frequency over a period of time.

The binomial distribution is one of the most common distributions used in industry when dealing with attributes or discrete measures. Among the most common applications is the attribute control chart covered in more detail in Chapter Five. For now, consider this simplified example where a product line is producing grommets for cables and wires to go between the firewall and the passenger compartment in an automobile. Several characteristics of this product are quality measures such as outside diameter, inside diameter, inside diameter of contact ridge, roundness, flatness, absence of burrs, blemishes, and bubbles. Many of those are continuous measures, but some are attributes (blemishes, bubbles, and burrs). Although the others are continuous, they can be treated as attributes. As the product comes off the line, inspectors or cameras can detect non-conformities (including those that are continuous). If the product shows any non-conformity, the whole piece is rejected as non-conforming; this gives an individual probability per piece of being conforming or non-conforming as P(non-conforming) = 0.5. Over the last several years of collecting data on this line, the defect rate is known to be 0.2 — obviously a line in need of improvement. This defect rate represents what is known in discrete statistics as an expected value (similar to a mean or average).

The experiment is designed to collect 10 items from the constant flow of product every 30 minutes from three 8-hour shifts for the next three days. This experiment will provide 1440 data for analysis and construction of an attribute control chart. Knowing that the defect rate is P(defect) = 0.2, what is the probability distribution of non-conforming pieces during the trial? Using equation 3.5:

$P(0) = 10! / 0!(10-0)! \, (.2^0 \, (1-.2)^{10-0}) = 0.107374182$
$P(1) = 10! / 1!(10-1)! \, (.2^1 \, (1-.2)^{10-1}) = 0.268435456$
$P(2) = 10! / 2!(10-2)! \, (.2^2 \, (1-.2)^{10-2}) = 0.301989888$
$P(3) = 10! / 3!(10-3)! \, (.2^3 \, (1-.2)^{10-3}) = 0.201326592$
$P(4) = 10! / 4!(10-4)! \, (.2^4 \, (1-.2)^{10-4}) = 0.088080384$
$P(5) = 10! / 5!(10-5)! \, (.2^5 \, (1-.2)^{10-5}) = 0.026424115$

$P(6) = 10! / 6!(10 - 6)! (.2^6 (1 - .2)^{10-6}) = 0.005505024$
$P(7) = 10! / 7!(10 - 7)! (.2^7 (1 - .2)^{10-7}) = 0.000786432$
$P(8) = 10! / 8!(10 - 8)! (.2^8 (1 - .2)^{10-8}) = 7.3728E-05$
$P(9) = 10! / 9!(10 - 9)! (.2^9 (1 - .2)^{10-9}) = 4.096E-06$
$P(10) = 10! / 10!(10 - 10)! (.2^{10} (1 - .2)^{10-10}) = 1.024E-07$

A graph of the resulting distribution is shown in Figure 3.3 (first series of graph bars or P (defect) = 0.2). Realize P(defect) = 0.2 is very high for a mass-produced product. After a few focused Six Sigma projects (see Chapter Eleven) affecting some quality improvements, the new, more realistic probability of selecting a defect is P(defect) = 0.02 shown in the second series of graph bars in Figure 3.3. Note the difference in the shape of the distribution made by changing only the known probability of non-conforming objects in the population.

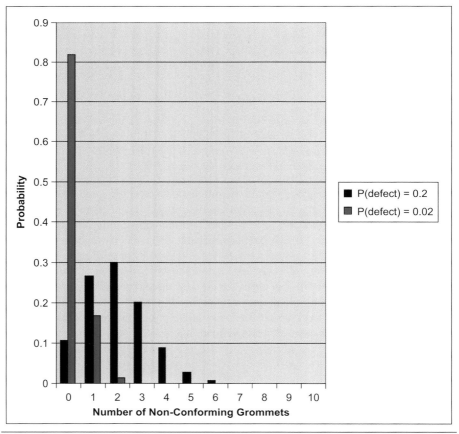

FIGURE 3.3  Binomial distribution of grommets

## The Poisson Distribution

For practical purposes the Poisson distribution is very similar to the binomial distribution. In fact, one can successfully approximate the other in many cases. The difference is the time element. A Poisson distribution is a discrete distribution showing probabilities of successes (or non-conformities in industrial terms) over a given time period; for example, the number of defects per hour or per shift. By contrast, the binomial distribution establishes probabilities in regard to an infinite number of items regardless of time. In industry, this time element can be redirected to objects such as cars and planes where it is important to ascertain the number of scratches and other blemishes per car, or the number of loose rivets in the fuselage per plane. Again, the assumptions remain the same. As with the binomial distribution, the Poisson distribution is determined from an infinite population. The measure is discrete (conforming or non-conforming), giving an individual probability of success as P(non-conforming) = 0.5, and trials are independent.

The formula for the Poisson distribution is as follows:

$$P(\# \text{ non-conforming}) = e^{-\lambda} \lambda^k / k! \qquad (3.6)$$

where:

  $e$ = base of natural logarithm = 2.718281
  $k$ = nonconforming objects drawn during the trial
  $\lambda$ = known mean rate, or average occurrence (of non-conformities) during the time period (similar to p in the binomial distribution and typically designated by the Greek letter lambda)

In industry, the formula is frequently represented as:

$$P(\# \text{ non-conforming}) = (np)^k / k! \; e^{-np} \qquad (3.7)$$

substituting np for $\lambda$ where n is number in the sample and p is the known probability (expected value) of non-conformities in the population.

Suppose a process in a manufacturing setting requires treating a metallic surface with a chemical bath to prepare the surface for further operations. The treatment is periodic and intermittent with limited or no ability to determine when the bath is required. The chemical solution is expensive and has a useful shelf life of 6 days. If the bath is required and the solution is not available, the process must be shut down until the solution arrives. Shutting down the process becomes very expensive. Management determines subjectively that between the cost of shortage and the cost of wasting unused solution, the best service level should be no more than 80%. In other words, they are willing to take a 20% risk of running out of the solution so they do not have to stock more of the solution. Over a long period of time it is estimated the bath is required twice within the 6-day period. What is the probability of no bath being required, 1 bath being required, 2 baths, and so on?

The calculated values of this distribution are:

  $P(0) = 2.718281^{-2} \; 2^0 / 0! = 0.135335283$
  $P(1) = 2.718281^{-2} \; 2^1 / 1! = 0.270670566$

$P(2) = 2.718281^{-2}\, 2^2 / 2! = 0.270670566$
$P(3) = 2.718281^{-2}\, 2^3 / 3! = 0.180447044$
$P(4) = 2.718281^{-2}\, 2^4 / 4! = 0.090223522$
$P(5) = 2.718281^{-2}\, 2^5 / 5! = 0.036089409$
$P(6) = 2.718281^{-2}\, 2^6 / 6! = 0.012029803$
$P(7) = 2.718281^{-2}\, 2^7 / 7! = 0.003437087$
$P(8) = 2.718281^{-2}\, 2^8 / 8! = 0.000859272$

These calculated values create the distribution shown in Figure 3.4. The individual probability values can also be found in the standard Poisson distribution tables in Appendix A, also shown here in Table 3.1.

First find the table corresponding to a $\lambda = 2$. Then follow the probabilities respectively for the desired value of k. Note the second column of probabilities. This column shows cumulative probability values. It simply adds each of the individual values to a total of 1.00. This column is useful in following through with the example. The under-informed manager wanted no more than an 80% service level for stocking the

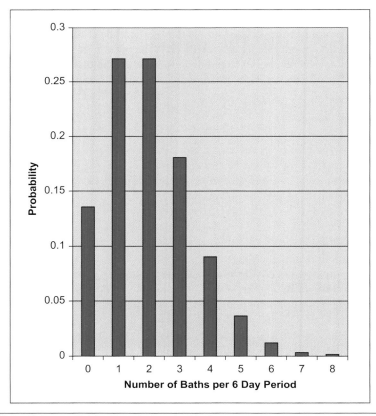

FIGURE 3.4  Poisson distribution of chemical baths

TABLE 3.1  Poisson values with $\lambda = 2$

| k | Individual Probability | Cumulative Probability |
|---|---|---|
| 0 | 0.135335283 | 0.135335283 |
| 1 | 0.270670566 | 0.406005850 |
| 2 | 0.270670566 | 0.676676416 |
| 3 | 0.180447044 | 0.857123460 |
| 4 | 0.090223522 | 0.947346983 |
| 5 | 0.036089409 | 0.983436392 |
| 6 | 0.012029803 | 0.995466194 |
| 7 | 0.003437087 | 0.998903281 |
| 8 | 0.000859272 | 0.999762553 |

chemical solution. Under the Cumulative Probability column, locate where 0.80 fits. Because of the discrete character of the distribution, it appears to be between 0.677 and 0.857. Those probabilities correspond to stocking levels of 2 or 3. Because stocking 3 units of solution increases the service level to 85.7%, the stocking level is dropped to 2 units of solution and the actual service level becomes 67.6%. Management may want to reconsider the service level and increase stocking levels to 3 units of solution.

Incidentally, the Poisson distribution was developed by and named after Simeon Poisson (1781–1840). He disseminated this probability research two years before his death in his 1838 publication, *Research on the Probability of Judgments in Criminal and Civil Matters*. He was a mathematician, physicist, and astronomer contributing to statistics, electromagnetism, the planetary theory, and much more.

The three distributions — hypergeometric, binomial, and Poisson — have similarities; their difference is primarily size of population. The higher the population, the greater the sample size can be. As the sample size continues to grow, the distribution becomes smoother. Ultimately, even a discrete distribution can be approximated by a continuous distribution. Continuous distributions are covered in Chapter 4, Basic Statistics for Continuous Distributions.

## Chapter Three Discussion

1. Distinguish among classical, relative, and subjective probabilities.
2. Distinguish among a trial, an event, and an outcome.
3. Describe the difference between mutually exclusive and non-mutually exclusive events.
4. Describe the difference between independent and dependent events.
5. Who is going to win the World Series this year? Explain the types of probability you used to arrive at your answer.

6. What is meant by the addition rule? What is meant by the multiplicative rule?
7. What are the characteristics of a discrete distribution?
8. In what case would the hypergeometric distribution be used?
9. What is the difference between the Binomial distribution and the Poisson distribution? When would each be used?

## Chapter Three Problems

1. Use a coin, a deck of cards, and a die to test the examples given in this chapter.
2. What is the probability of drawing a 7 (of any suit) out of a deck of fair cards? What probability does this exemplify?
3. Out of 100 units tested in an industrial process, 4 failed. What is the failure rate of the process? Given this failure rate, what is the expected number out of 850 units?
4. What is the difference between a combination and a permutation? Give examples of each.
5. Ten cards are numbered 1 through 10 and placed in a box. What is the probability of drawing the sequence 3, 6, and 7 when the card is replaced after each trial? What is the probability of drawing the sequence 3, 6, and 7 when the card is not replaced after each trial?
6. Calculate the hypergeometric probabilities of the exercise in problem 5.
   Add a condition where a particular suit is disqualifying and recalculate the probabilities. Try to graph this distribution in Minitab® or Excel®.

# FOUR
## Basic Statistics for Continuous Distributions

Chapter Three discussed discrete measures, such as attributes, that resulted in whole numbers. The distributions they generated were discrete distributions that provided a picture of the shape and behavior of the collected data. This chapter follows that same logic; however, the measures are variable — what is known as continuous. In each case, the calculated probabilities show an analytic version of the distribution, with the charts showing a graphic picture of that distribution. This branch of statistics is called *descriptive statistics*; it is used simply to show shape, key measures, and general characteristics of the distribution.

As with discrete distributions, one purpose of continuous distributions is to compare the sample distribution with that of a standard distribution. In discrete distributions, standard distributions included the binomial and the Poisson distribution. In continuous distributions, this standard distribution is called the *normal distribution*. The normal distribution is a continuous theoretical distribution for an infinite number of samples. Therefore, the higher the sample size, the closer the sample distribution resembles a normal distribution (if the population is naturally normal), given the randomness and non-bias approaches in data collection discussed in Chapter Two.

There are two measures of a sample in terms of descriptive statistics: measures of central tendency, and measures of dispersion. As a reminder, descriptive statistics simply show characteristics of a distribution. With certain measures, the researcher or operations technician can understand and describe how a set of data behaves. At this point, descriptive statistics would not indicate any behavior outside of the sample such as to a population. Basic inferential statistics, which help us infer such behavior, will be covered in Chapter Nine.

## Measures of Central Tendency

Central tendency is the phenomenon of data collecting around the center of the sample or distribution. As with a normal distribution, these data would ideally be centered and symmetrical. Sample distributions that closely resemble these normal distributions allow a more valid comparison with the properties of the normal distribution. There are three common measures of how data collects toward the center: the mean, the median, and the mode. When the distribution is normal, these three measures are equal.

### Mean

The mean is the arithmetic average as shown by the following equation from Chapter Two:

$$\bar{x} = \Sigma x/n \tag{2.1}$$

where:

$\Sigma x$ = summation of all data in the sample
$n$ = the number of data in the sample

Suppose we collect data measuring the time it takes (in seconds) for a welder operator to weld a distance of eight inches on 0.25 inch steel plate. The data is tabulated in Table 4.1.
Note that

$$\Sigma x = 198$$
$$n = 5$$

Plugging these values into equation 2.1 gives:

$$\bar{x} = 198/5 = 39.6$$

Therefore it takes an average or mean time of 39.6 seconds for this operator to weld a distance of eight inches. Note that 39.6 is not rounded to 40. This is because the mean is a continuous measure (time) and there is no need to represent it in whole numbers. In fact, rounding this measure to a whole number makes the mean incorrect. Rounding can

TABLE 4.1  Time to Weld in Seconds

| n | x |
|---|---|
| 1 | 38 |
| 2 | 40 |
| 3 | 37 |
| 4 | 42 |
| 5 | 41 |
| Σ | 198 |

be used in cases where the accuracy of the decimal place becomes insignificant or if the mean delivers an irrational number. For example, if the mean is 39.666666666...., and the significant decimal placement is thousandths of a unit, this measure is rounded to 39.667.

Each datum is treated the same in this example, which applies to any continuous distributions no matter how many data are in the sample. There are cases where each datum is treated differently because of importance or weight. This is referred to a *weighted mean*. Consider a professor administering four tests during a semester at college. The first three tests are each worth 20% of the student's final grade and the fourth test (the final) is worth 40%. Equation 2.1 is now modified to consider this difference in weight, and the weighted mean becomes:

$$\bar{x}_w = \Sigma xw / \Sigma w \qquad (4.1)$$

where:

$$w = \text{weight in decimal equivalent}$$

For example, a student receiving 84, 92, 88, and 82 on each respective test would receive a final grade of:

$$\bar{x}_w = [(84 \times .2) + (92 \times .2) + (88 \times .2) + (82 \times .4)] / 1 = 87.6$$

This modification is similar to another form of mean called a *frequency mean* where data are multiplied by the frequency of occurrence of each value such as:

$$\bar{x}_f = \Sigma xf / \Sigma f \qquad (4.2)$$

where:

$$f = \text{number of reoccurring times (frequency)}$$

This type of mean is most useful when there are larger numbers of data where there will be more than one datum holding the same value. If the welder in the previous example improves the process and welds 50 additional samples, the times (again in seconds) may start to reoccur. Suppose the tabulated values are those shown in Table 4.2. In this case:

$$\Sigma f = 50$$

TABLE 4.2  Time to Weld in Seconds (Frequency Mean)

| x | f |
|---|---|
| 31 | 8 |
| 32 | 12 |
| 33 | 15 |
| 34 | 13 |
| 35 | 2 |
| Σ | 50 |

Eight times the welder takes 31 seconds to weld the steel. Twelve times the welder takes 32 seconds to weld the steel, and so on. This frequency mean computes as:

$$\bar{x}_f = [(31 \times 8) + (32 \times 12) + (33 \times 15) + (34 \times 13) + (35 \times 2)] / 50$$
$$= (248 + 384 + 495 + 442 + 70) / 50 = 32.78$$

Note how each datum is multiplied by each respective frequency. The frequencies are similar to the weights in the weighted mean (equation 4.1). The frequency mean is also used when data are grouped as seen in Chapter Two. Recall the example regarding the ppm contamination of chloride in the pond near a manufacturing plant. Cell midpoints become the value (x) and the frequency is the number of data within each cell. For the data in Figure 2.5, the mean is:

$$\bar{x}_f = [(12 \times 7) + (17 \times 11) + (22 \times 18) + (27 \times 14) + (32 \times 9)] / 59$$
$$= (84 + 187 + 396 + 378 + 288) / 59 = 22.593$$

Note that 12, 17, 22 and so on are the cell midpoints. In turn, 7, 11, 18 and so on represent the number of data in each of the respective cells. The total number of data in that study is:

$$n = 59$$

## Median

The median is the middle value of the distribution. In a perfectly normal distribution, the median would equal the mean. However, even a sample close to normal tends to vary, and the result is a difference between the median and the mean. The median is useful when considering measures such as percentiles — it is the 50% percentile. It is meaningful when distributions are not symmetrical, for example, when data tend to congregate to one side or the other in the distribution.

Because the median is the middle of the data, it can help to reveal the frequency of data below that point (on the negative, or left side of the distribution) and the frequency of data above that point (on the positive, or right side of the distribution). In a case where the greater frequency of data is above the median (on the positive, or right side of the distribution), the mean tends to move away from the median toward the negative (left) side of the distribution (skewed to the left or negatively skewed), as shown in Figure 4.1. In a case where the greater frequency of data is below the median (on the negative, or left side of the distribution), the mean tends to move away from the median toward the positive (right) side of the distribution (skewed to the right or positively skewed), as shown in Figure 4.2.

The median is calculated by finding the middle number in the sample. First arrange the data in an array. If the sample contains an odd number of data, the median is a real datum, residing exactly in the middle. For example, if n = 45, the median is 45/2 = 22.5, rounded to the 23rd number in the sample array. Twenty-two numbers are below that value and twenty-two numbers are above it. If the sample contains an even number of data, the median becomes the mean of the two middle numbers. For example, if there are 46 numbers in an array, 46/2 = 23, and the median is the simple mean of

# Basic Statistics for Continuous Distributions 67

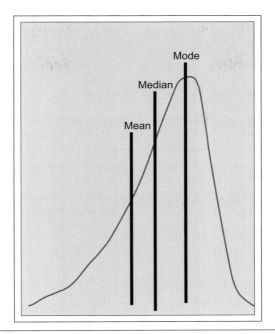

FIGURE 4.1   Negative skew

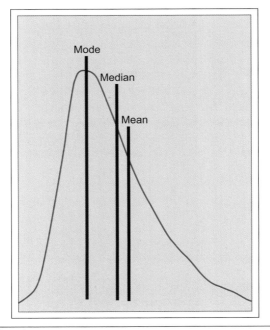

FIGURE 4.2   Positive skew

the 23rd and 24th numbers. Suppose the 23rd number is 56 and the 24th number is 58. The median is (56 + 58) / 2 = 57.

Grouped data present a slightly more complicated equation, but the idea is exactly the same. Consider the PPM example in Chapter Two (Figure 2.8) with 120 data grouped into 11 cells, represented here again in Figure 4.3 (see Appendix E). The medium is calculated using the following equation:

$$M_g = L_{cm} + [(.5n - cf_i) / f_i] \, i \qquad (4.3)$$

where:

$L_{cm}$ = lower boundary of the cell containing the median
$i$ = interval width
$cf_i$ = cumulative frequency of data below the cell containing the median
$f_i$ = frequency of data within the cell containing the median

Note: the equation calls for identifying the cell that contains the median. Note also the median cell has a cumulative frequency equal to or greater than n/2.

First, find the median cell. For the PPM example, the exact half of the sample distribution is .5n = 60. Then, starting from the first cell on the left, the cumulative frequency

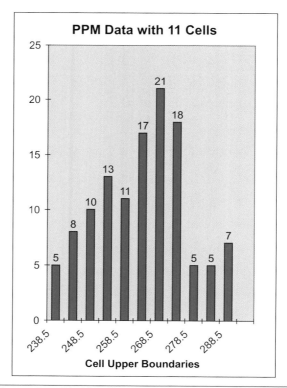

FIGURE 4.3   PPM grouping with 11 cells

is the running sum of all cells as 5 (cell one), 13 (5 + 8) (cell two), 23 (13 + 10) (cell three), 36 (23 + 13) (cell four), 47 (36 + 11) (cell five), and 64 (47 + 17) (cell six). There is no need to add further because sixty (n/2) is greater than 47 and less than 64. The median is therefore contained in the sixth cell. Applying equation 4.3:

$$M_g = L_{cm} + [(.5n - cf_i) / f_i] I = 258.5 + [(60 - 47) / 17] 5 = 262.3235$$

This result makes perfect sense because of the shape of the distribution. Recall the mean is 260.5 for this distribution and the shape of the distribution is negatively skewed, with more data on the right. The median is the point for which half the data is above and half the data is below. In this case, the mean is found toward the negative side of the median (left). Therefore, Figure 4.4 shows this distribution as skewed slightly negative (left).

(Incidentally, the ungrouped median of this distribution is actually at 261.5. The curve in Figure 4.4 appears more negatively skewed than it actually is. The little bump in the data around the 251 datum actually pulls the central tendency slightly toward the left side of the distribution.)

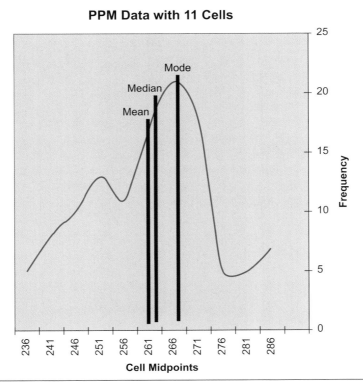

FIGURE 4.4   Mean and medium for PPM data

## Mode

The mode is the point where most data fall. It represents the most common frequency. In a normal distribution, the mean, median, and mode share the same value. In cases where the distribution is skewed, the mode moves farther away from the skew than the mean and the median; it follows the highest point in the distribution (Figure 4.1). In some cases, distributions have more than one mode. Consider again the PPM data in Figure 4.3. The predominate mode falls in the 7th cell containing 21 data. However, the shape shows another grouping of data around the 4th cell containing 13 data. This is a bi-modal condition. In ungrouped data, the mode is the most frequent number. If these PPM data were ungrouped there would be several modes — a condition called *multi-modal*. Mode is used not so much as a quantitative measure, but an observation of the most common frequency.

The mode creates an interesting comparison with that of the mean in skewed distributions; it is the most probable value if subjects are chosen randomly with no bias. The mode is the most likely scenario.

By themselves, the mean, median, and mode do not adequately describe distributions. Unfortunately, a common mistake in attempting to describe a distribution is reporting a measure of central tendency (usually the mean) with no other information. These measures of central tendency taken alone describe little. Imagine a group of people where the mean age is 50. Initially, the reader may develop an idea about the group and the age of the people in the group. This is an attempt to describe the distribution. A little more information can completely change this perspective. There are four people in the group. Two of them are 90 years old, and two of them are 10 years old. None of them are actually 50. The mean of 50 (along with other measures of central tendencies) makes no sense without an additional knowledge of the distribution, namely dispersion. Even knowing the dispersion (a range of 80 years) and the mean doesn't fully describe this distribution. Because two subjects are 10 years old and two subjects are 90 years old out of a total of four subjects, this distribution is bi-modal. In addition, note that this distribution is symmetrical where the mean and the median are equal. In reality, the chances of a non-biased, randomly collected sample developing into a distribution as the one described here is almost nil. Nevertheless, caution must be taken when attempting to describe any distribution with too little information.

## Measures of Dispersion

Dispersion reveals how spread-out the data are. Not only can it describe how far the data spread out, certain measures can determine the probability of a particular datum (or range of data) existing anywhere within the distribution. Dispersion serves to better describe the actual shape of the distribution. Toward the end of this chapter, there will be a brief discussion on some of the more common shapes a distribution may display because of dispersion. For now, look at the most common statistics of these dispersion measures. There are three common measures of dispersion: the range, the variance, and the standard deviation.

## Range

Range is the difference between the highest datum and the lowest datum. It is the total spread of the data. This is the simplest description of dispersion; as such, its use is limited in much of statistics. The range alone cannot be used to give a good description of the actual shape of the distribution. It can only show the distribution's spread. Mathematically the range is given by:

$$R = x_h - x_l \tag{4.4}$$

where:

$x_h$ = highest datum
$x_l$ = lowest datum

Although the range is only a simple measure of dispersion, it is used frequently in industry with the use of quality control (QC) charts (see Chapter Six). Recall that continuous distribution uses variable data. The most common QC chart for variable data is the "X-bar($\bar{X}$) and R" chart. Here, $\bar{X}$ symbolizes the mean and R symbolizes the range. Together, these two measures allow a view of a particular process' quality in progress, providing information as a means to understand, analyze, describe, and correct quality issues.

For the PPM data given in Table 2.6 (see Appendix E), the lowest datum is 234 and the highest datum is 287. The range is R = 287 − 234 = 53. This method is applied to both ungrouped and grouped data.

## Variance and Standard Deviation

Variance is defined as the average of the squared differences of each datum from the mean. These are squared simply to make positive any negative data that may exist. Unlike the range, which uses only two data to calculate dispersion, the variance utilizes all the data in relation to the mean and allows calculations of probabilities at any point under the distribution. It gives a better picture of the shape of the distribution rather than just the spread. The standard deviation is the square root of the variance.

Think of a simple data set consisting of 4, 5, and 6. The mean of these data is 5. The difference between 4 and 5 is −1, the difference between 5 and 5 is 0, and the difference between 6 and 5 is 1. Squaring −1 gives 1, 0 squared is 0, and 1 squared is 1. Summing the squares gives 1 + 0 + 1 = 2. Computing the average of these squared differences determines the variance. There are 3 data, so,

$$\Sigma x/n = 2/3 = 0.667$$

The standard deviation is the square root of the variance or 0.817.

Larger sets of data are treated the same way; however, they may require some sort of table or computer program to keep track of all the numbers and computations. Consider the data listed in Table 4.3. They have been arrayed in an ascending manner; the square of the difference between each datum and the mean has been calculated where the mean is 20.933 (equation 2.1).

## 72   Chapter Four

TABLE 4.3   Working with Larger Sets of Data

| x | Differences from mean | Colomn 2 values squared |
|---|---|---|
| 15 | -5.93333 | 35.2044 |
| 16 | -4.93333 | 24.33774 |
| 16 | -4.93333 | 24.33774 |
| 17 | -3.93333 | 15.47108 |
| 18 | -2.93333 | 8.604425 |
| 20 | -0.93333 | 0.871105 |
| 21 | 0.06667 | 0.004445 |
| 21 | 0.06667 | 0.004445 |
| 22 | 1.06667 | 1.137785 |
| 24 | 3.06667 | 9.404465 |
| 24 | 3.06667 | 9.404465 |
| 24 | 3.06667 | 9.404465 |
| 25 | 4.06667 | 16.5378 |
| 25 | 4.06667 | 16.5378 |
| 26 | 5.06667 | 25.67114 |
| Σ |  | 196.9333 |

Working with these data,

$$\Sigma = 196.9333$$

Following through, the variance of this distribution is the mean of column 3 or

$$\Sigma x/n = 196.933/15 = 13.129$$

The standard deviation is the square root of $\Sigma x/n$, or 3.623. The variance is the square of the standard deviation; therefore, the equation for the standard deviation is as follows:

$$s = \sqrt{\Sigma(x - \bar{x})^2 / (n-1)} \qquad (4.5)$$

One variation of this equation is easier to use with calculators and is as follows:

$$s = \sqrt{[n\Sigma x^2 - (\Sigma x)^2]/[n(n-1)]} \qquad (4.6)$$

Note the important difference between the summation of x squared, and the square of the sum of x.

Equation 4.6 requires only the x column, a column of x square, and the sums of each of those columns. Table 4.3 can be rebuilt as Table 4.4.

## Basic Statistics for Continuous Distributions

TABLE 4.4  A Rebuilding of Table 4.3

| | x | x squared |
|---|---|---|
| | 15 | 225 |
| | 16 | 256 |
| | 16 | 256 |
| | 17 | 289 |
| | 18 | 324 |
| | 20 | 400 |
| | 21 | 441 |
| | 21 | 441 |
| | 22 | 484 |
| | 24 | 576 |
| | 24 | 576 |
| | 24 | 576 |
| | 25 | 625 |
| | 25 | 625 |
| | 26 | 676 |
| Σ | 314 | 6770 |

Working with these data,

$$\Sigma x = 314$$
$$\Sigma x^2 = 6770$$

Now, plugging in equation 4.6:

$$s = \sqrt{[n\Sigma x^2 - (\Sigma x)^2]/[n(n-1)]}$$

$$= \sqrt{(15)6770 - (314)^2 / 15(15-1)} = \sqrt{101550 - 98596 / 210} = \sqrt{14.067} = 3.75$$

With only slight differences attributed to rounding numbers, there is little difference between using equation 4.6 and using the calculations for Table 4.3; however, equation 4.6 is much simpler and easier to manage.

## 74   Chapter Four

For grouped data, this formula changes slightly to:

$$s = \sqrt{[n\Sigma(fx^2) - (\Sigma fx)^2]/[n(n-1)]} \quad (4.7)$$

where:

$$f = \text{frequency of data within each cell}$$

Suppose the example above was in regard to Rockwell hardness using the C scale (RC) in determining the hardness of a ceramic/metallic compound. Each test piece, or coupon, was inserted into the measuring instrument and the diamond tip of the tester was pressed into the material. Each coupon was tested five times and the average of those five readings per coupon was recorded as one datum. The experiment was repeated, generating 178 data (see Appendix E), and the data were grouped into 15 cells. Midpoints of those cells started with 15 and progressed through 29. Frequencies within each of the cells and the necessary computations are given in Table 4.5.

Working with these data,

$$\Sigma f = 178$$
$$\Sigma fx = 3916$$
$$\Sigma fx^2 = 87242$$

TABLE 4.5   Working with Grouped Data

| cell | cell midpoint | f | $x^2$ | fx | $fx^2$ |
|---|---|---|---|---|---|
| 1 | 15 | 1 | 225 | 15 | 225 |
| 2 | 16 | 2 | 256 | 32 | 512 |
| 3 | 17 | 4 | 289 | 68 | 1156 |
| 4 | 18 | 7 | 324 | 126 | 2268 |
| 5 | 19 | 12 | 361 | 228 | 4332 |
| 6 | 20 | 19 | 400 | 380 | 7600 |
| 7 | 21 | 28 | 441 | 588 | 12348 |
| 8 | 22 | 32 | 484 | 704 | 15488 |
| 9 | 23 | 28 | 529 | 644 | 14812 |
| 10 | 24 | 19 | 576 | 456 | 10944 |
| 11 | 25 | 12 | 625 | 300 | 7500 |
| 12 | 26 | 7 | 676 | 182 | 4732 |
| 13 | 27 | 4 | 729 | 108 | 2916 |
| 14 | 28 | 2 | 784 | 56 | 1568 |
| 15 | 29 | 1 | 841 | 29 | 841 |
| Σ | | 178 | | 3916 | 87242 |

Using the standard deviation formula for grouped data (equation 4.7), the standard deviation is computed as follows:

$$s = \sqrt{[n\Sigma(fx^2) - (\Sigma fx)^2]/[n(n-1)]}$$
$$= \sqrt{[178(87242) - 3916^2]/[178(178-1)]}$$
$$= \sqrt{(15529076 - 15335056)/31506} = \sqrt{6.158192} = 2.48157$$

The mean (using equation 4.2) is as follows:

$$\bar{x}_f = \Sigma xf / \Sigma f = 3916/178 = 22$$

With the mean, standard deviation, and some basic assumptions regarding normality, the distribution can be described and graphed as a curve — filling in the numbers at the bottom of the graph over the *x*-axis (Figure 4.5). The standard deviation was used to determine these points along the bottom of the graph. They were added to or subtracted from the computed mean which lies in the middle. Note there are six segments along the *x*-axis of the graph: three above the mean and three below. Each was calculated using the standard deviation. These segments are significant because they represent areas corresponding to the empirical rule of the normal distribution (see the next section of this chapter). The standard deviation shows spread, as does the range. Unlike the range, the standard deviation (and hence variance) gives shape to the description of the distribution.

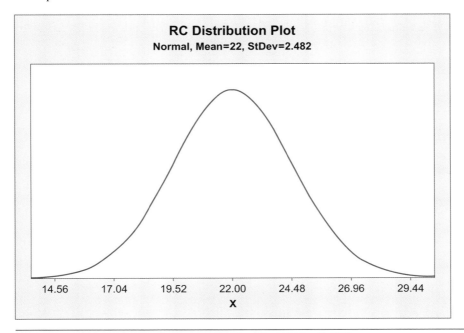

FIGURE 4.5  RC distribution based on standard deviation = 2.482

## The Empirical Rule

The distribution seen in Figure 4.6 is a sample distribution. If the frequency in the study is very large (in the thousands perhaps), the sample shape will be close to the population. A normal distribution represents not only a perfect population, but also a perfect sample if there are a high enough number of data (n). A normal distribution will have certain characteristics regarding height-to-width proportions along the spread (in theory, rarely in practice). These proportions are determined by what is called the *empirical rule*, where empirical refers to the standard established by experience. The empirical rule helps map out areas of the normal distribution as they relate to variance — or more specifically, standard deviation (Figure 4.6). Because the normal distribution depicts a perfect population, terms such as μ (mu: population mean) and σ (sigma: population standard deviation) are used instead of $\bar{x}$ (sample mean) and s (sample standard deviation).

The empirical rule is sometimes referred to as the 68-95-99.7 rule. It means that 68.26 percent of the area under the normal distribution resides within one standard deviation of the mean (above and below the mean). In turn, 95.46 percent of the area resides within two standard deviations (above and below the mean) and 99.73 percent of the area resides within three standard deviations (above and below the mean). Less than 0.3 percent of the area resides outside of three standard deviations from the mean.

This area within the curve is directly related to probability. Superimpose a sample distribution over a normal distribution. Determining the area below a selected part of the normal distribution provides the probability that a randomly selected non-bias subject in the sample distribution has that value. Think of the normal curve as a template. Holding the sample curve against that normal curve will show to what extent

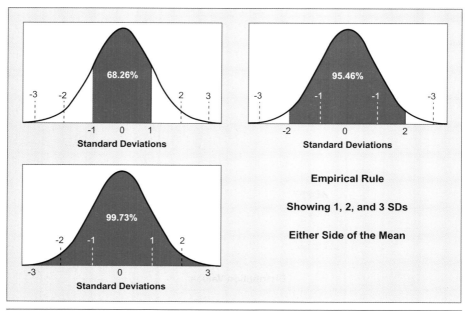

FIGURE 4.6  Normal distribution showing empirical rule

the sample deviates from normal. If the sample curve is close to normal, the normal curve's properties can be used for the sample curve. In this case, the probability of any position within the distribution can be determined.

But describing the distribution this way is contingent upon the distribution being close to normal. If the distribution deviates too far from normal, this method will not apply (outside of the template). In statistics, this means the assumption of normality is not met. As discussed in Chapter Three, basic assumptions for normality include independence, randomization, and an adequate sample size compared to the population (high n).

Refer to Figures 4.5 and 4.6. Continuing with the Rockwell hardness example, the mean is RC22. It is easy to understand that 50% of the data fall below the mean and 50% of the data fall above the mean. Therefore, the probability of a randomly selected coupon reading being above or below RC22 is 0.50. Subsequent probabilities are calculated with the same idea.

Applying the empirical rule, 68.26% of the data are within one standard deviation of the mean. Comparing the normal distribution to the sample distribution, we can surmise 68.26% of hardness readings are between RC19.52 and RC24.48 (mean minus one standard deviation and mean plus one standard deviation). Half of 68.26% is 34.13%. Therefore, on the negative side, 34.13% of the readings would be between the mean and RC19.52.

In turn, approximately 95.5% of the data fall within two standard deviations. Subtracting 68.26% from 95.46% gives 27.2%, the area covered by the second standard deviation. Half of 27.2% of the readings, or 13.6%, are between RC17.04 and RC19.52 on the negative side of the curve. In other words, a randomly selected hardness reading has a 0.136 probability of existing between RC17.04 and RC19.52. The shaded area in Figure 4.7

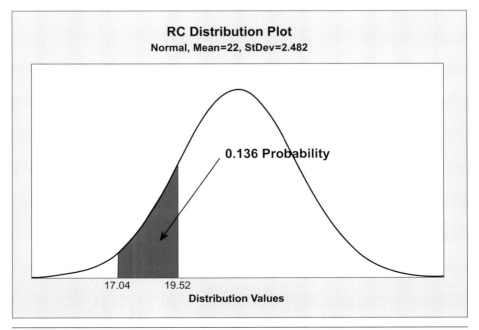

FIGURE 4.7  Area between 17.04 and 19.52 representing 0.136 probability

represents this area of the curve. The same pattern is true on the positive side of the mean, between RC 24.48 and RC26.96.

Given this material's intended use, suppose the lower limit of hardness (too soft) is RC17.04, located two standard deviations below the mean. What is the probability of a randomly selected sample of this material being below this useful limit, or too soft to use? The area below the mean is half (50%) of the total area. The area of two standard deviations is 95.46%. Of that, the half below the mean is 47.73%. Therefore, (50% − 47.43%) = 2.27% is the area below the second standard deviation. There is a 0.0227 probability that a randomly selected piece of this material is below the useful limit of RC17.04.

## The Standard Normal Z-Distribution

Using whole standard deviations is fine for describing inclusive areas of the distribution, but what about specific locations on the distribution? The Z-distribution isolates those areas or probabilities or perhaps just one point along the distribution. It displays a point in the distribution's area that is either negative or positive and can be a fraction as well.

For example, a specific point can exist 2.33 standard deviations above the mean. The Z-distribution calibrates the mean of the sample to zero. It then signifies three standard deviations above the mean as 1, 2, and 3, and signifies three standard deviations below the mean as −1, −2, and −3 (Figure 4.8). Next, assuming a normal sample

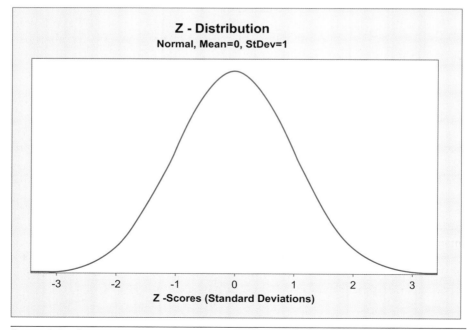

FIGURE 4.8   Z-distribution

distribution, sample standard deviations can be transformed into z-scores that can utilize the Z-distribution to determine the area under the curve or probability. The Z-distribution becomes sort of a scale with which to measure the sample distribution. Sometimes this process is referred to as *standardization*.

Because the distribution is continuous, the number of available z-scores is infinite and can be determined with most scientific calculators and spreadsheets. In practice, however, selected z-scores are identified in normal distribution tables where the corresponding area is tabulated into probability (Appendix B). Before these tables can be used, the sample standard deviation must be transformed into the z-score. This z-score transformation formula is:

$$z = x - (\mu / \sigma) \tag{4.8}$$

where:

$x$ = the value of interest
$\mu$ = the population mean
$\sigma$ = the population standard deviation

Note the use of population parameters ($\mu$ and $\sigma$) rather than sample statistics ($\bar{x}$ and $s$). Because the Rockwell hardness distribution has a relatively large n (178) and the shape of the distribution is symmetrical and close to a normal distribution, the sample is determined to be normal and represent the population. In industry, in practice, this is the most common method of meeting the assumption of normality. Several formal tests are introduced later in this chapter.

Putting the z-score to the test with the Rockwell hardness example, it was determined earlier 2.27% of the data fell below the RC17.04 reading (.0227 probability). Using equation 4.8 should return the same results:

$$z = x - (\mu / \sigma)$$
$$z = 17.04 - (22 / 2.48157) = -1.99873$$

This value (rounded to −2) indicates two standard deviations below the mean — exactly as calculated before. Looking now at the tabulated areas under the normal distribution in Appendix B, find −2 in the first column of the table and 0.00 in the top row. Follow these lines over and down into the area values until they converge. The area reported is 0.0228 or 2.28%. Again, this is the same as before with some rounding effect.

Now, suppose this company needs to know how much product falls above the upper specification limit of the material. In other words, the material is too hard. This information is required in part to increase the quality of the process, and to determine costs associated with scrap, because anything above or below limits is wasted. Engineers have determined the upper limit for this material given its intended use is RC27.33. Anything above that (too hard) must be scrapped. What is the probability of producing material above the upper limit? Applying equation 4.8:

$$z = x - (\mu / \sigma)$$
$$= 27.33 - (22 / 2.48157) = 2.147834$$

This amount is rounded to z = 2.15. Referring again to Appendix B, follow the first column down to 2.1 on the positive side. Then follow the top row over to 0.05. Where those lines converge shows the probability of 0.9842. In other words, 98.42% of the distribution's area falls below this value. The probability is 1.0 − 0.9842 = 0.0158 of randomly selecting a piece of material that is too hard or, in this case, producing scrap above the upper limit (Figure 4.9). Had z = 2.147834 not been rounded to z = 2.15, the exact probability would be 0.984137, giving a slightly higher probability over that point of 1.0 − 0.984137 = 0.015863.

Not all z-score tables are the same. One common variation shows total area under the curve (as represented in Appendix B). Another type shows scores from only half the curve. In reality, half the curve is all that is required because the standard normal distribution is perfectly symmetrical. Any probability can be calculated with the understanding that z-scores transformed from the sample distribution can be either negative or positive.

The probabilities found in the table and with the computer or calculator can also serve to determine points on the distribution. Suppose the manufacturer in the ongoing example wants to ignore the engineers and brag about a 99% usage of all produced material (taken from the middle). Figure 4.10 indicates the RC hardness they will be sending to the customer. Subtracting 0.99 from the whole 1.0 equals 0.01. Half of that is 0.005. Hence, finding probabilities of 0.005 and 0.995 will return z-scores for those data values.

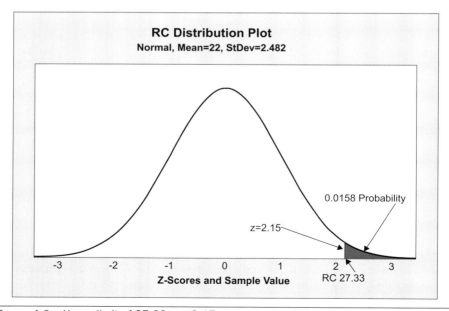

FIGURE 4.9   Upper limit of 27.33, z = 2.15

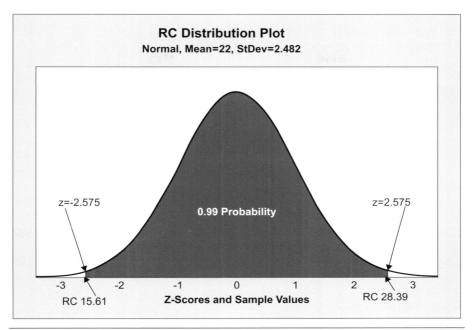

FIGURE 4.10   99% out of the middle

On the negative end, looking in the tables for area under the curve, the closest probabilities are 0.0049 and 0.0051 giving z-scores of −2.58 and −2.57. Splitting the difference between these z-scores gives a z-score of −2.575. Similarly, on the positive side, the closest probabilities are 0.9949 and 0.9951. Again, splitting the difference gives a z-score of 2.575. Because the standard normal distribution is symmetrical, you only need to find the first of these. The next step is to determine the RC values in the sample distribution corresponding to these z-score:

$$\text{Lower RC value} = \bar{x} - z(s)$$
$$= 22 - 2.575(2.48157) = RC15.61$$

and:

$$\text{Upper RC value} = \bar{x} + z(s)$$
$$= 22 + 2.575(2.48157) = RC28.39$$

For this example, the manufacturer is delivering a portion of material that is too hard on one end of the distribution and too soft on the other. Ignoring engineers and their specification limits does not typically occur in industry. However, this example shows the ability to work the procedure backwards and use interlinear interpolation (splitting the difference) between two probabilities (z-scores).

## Abnormally Shaped Distributions

As mentioned above, an assumption of normality must be met before attempting to describe samples with these statistics. Several tests are available for determining if a sample distribution is normal and, as such, can utilize the standard normal distribution for comparing and computing area. These include the Shapiro-Wilk test, the Kolmogorov-Smirnov test, the Anderson-Darling test, and the Cramér-von Mises test. Most of these tests make reference to the empirical rule and how well the tests fit into that characteristic of the normal distribution. These tests and Chi-square tests are often called *goodness of fit* tests. It is beyond the scope of this text to delve into these tests primarily because of their limited use in industry.

This chapter has already discussed the most common test for normality. The higher the sample size is, the more representative the sample is of the population. Keep a high n, but also plot and examine the distribution to see if it looks normal. Although this method may involve trial and error, it is the one most commonly used in industry. When you work with statistics on a regular basis, you must become familiar with the general shape of a normal distribution. Two more analytic tests used in industry offer a better idea of normality than just looking at the distribution. These tests consider the values of skewness and kurtosis.

### Skewness

Skewness is the lack of symmetry in the distribution. As skewness occurs, the measures of central tendency deviate from each other. A normal curve has a skew of 0, whereas a skewed distribution will deviate in a positive or negative direction. Figures 4.1 and 4.2 are examples of skew. For the ceramic/metallic material hardness example, the skew is also 0. This distribution may not be normal, but a skew of 0 does show the distribution is perfectly symmetrical — the mean, median, and mode are the same. Any deviation from symmetry becomes skewness. The formula for skewness is as follows:

$$\gamma^1 = [\Sigma(x - \mu)^3] / [(n - 1)\sigma^3] \qquad (4.9)$$

### Kurtosis

Kurtosis is a similar measure; it indicates how peaked or pointed the distribution is. A normal distribution has a kurtosis of 0. A common method of calculating kurtosis is known as the *excess kurtosis*. As the shape of a distribution flattens, the excess kurtosis becomes negative. As the shape of a distribution sharpens and becomes taller, the excess kurtosis becomes positive. The formula for excess kurtosis is as follows:

$$\gamma^2 = \{[\Sigma(x - \mu)^4] / [(n - 1)\sigma^4]\} - 3 \qquad (4.10)$$

Using equation 4.10 with the hardness study given above, the kurtosis is:

$$\gamma^2 = \{[\Sigma(x - \mu)^4] / [(n - 1)\sigma^4]\} - 3$$
$$= \{21{,}178 / [(177)37.9233]\} - 3 = 3.155 - 3 = .155$$

Basic Statistics for Continuous Distributions  **83**

The distribute of the hardness study sample seen in Figure 4.9 appears close to normal, but is actually slightly taller. This is supported by the computed excess kurtosis of .155. Samples sharing a similar kurtosis to a normal curve are called *mesokurtic* distributions. These have a computed excess kurtosis of 0; along with the skew of 0, they can be compared to normal distributions. Caution must be taken, however, because there are no foolproof short-cut methods to determine normality. Examples of kurtosis include *leptokurtic* and *platykurtic* distribution shapes.

Leptokurtic distributions are tightly gathered in the middle of the distribution and have positive excess kurtosis. They show a high or peaked point around the mean, median, and mode (Figure 4.11). Imagine the failure strength of wooden beams. The type of wood definitely has an influence on how the beam fractures. Maple beams for

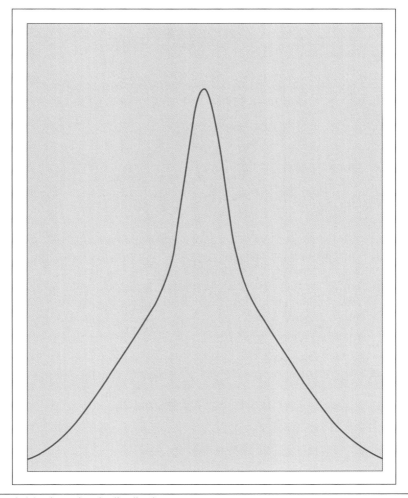

FIGURE 4.11   Leptokurtic distribution

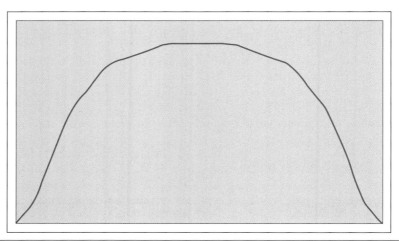

FIGURE 4.12  Platykurtic distribution

example tend to be strong to a certain degree, but when they do break, they fracture quickly. Figure 4.11 is a distribution shape that might occur with maple, for example.

Elm on the other hand is more resilient. Because it bends over and springs back, the fracture may occur over a wider area. Elm beams tend to be strong, but as they break, they give warning and fracture slowly. Figure 4.12 depicts a platykurtic distribution as may be expected from elm. Platykurtic distributions have negative excess kurtosis.

Both leptokurtic and platykurtic distributions are frequently symmetrical — the mean, median, and mode are equal, but the distributions do not have height and width characteristics allowing a reasonable comparison to the normal distribution. Discovering a sample is not normal is one thing. The more pertinent question is how do we correct the problem once we have determined a distribution is abnormal? How can we obtain a distribution where we can utilize standard features of a normal distribution? Again, sample size is critical. The higher the sample size is, the more expensive the experiment may be; however, the closer the sample size gets to the population, the better it will represent the population. A powerful technique commonly used to normalize a sample distribution is sampling by subgroups. This topic is covered in detail in Chapter Six regarding Statistical Process Control (SPC) with continuous measures.

## Chapter Four Discussion

1. What is the difference between an attribute and a variable?
2. What are the measures of central tendency and how are they used?
3. What are the measures of dispersion and how are they used?
4. How are the measures of central tendency and the measures of dispersion used together to describe a distribution?

5. Describe the empirical rule and explain how it is useful in statistics.
6. How does the normal curve relate to six standard deviations?
7. Describe z-scores, and explain how they are used in statistics.
8. What do the numbers in the z-tables mean?
9. What does interlinear interpolation mean?
10. Why would the z-distribution be referred to as a template, standard, or a no-go gauge?
11. Discuss some of the odd shape distributions can make. Explain the problem these shapes create and how statisticians can correct them.

## Chapter Four Problems

1. Using the data set provided for this problem in either Excel® or Minitab® (see Appendix E), determine the following:
   a. Mean
   b. Median
   c. Mode
   d. Variance and standard deviation
   e. Range
   f. Frequency diagram
2. Using the data set for Problem 1, group the data into intervals of 3, then recalculate the statistics shown above. Repeat the problem with intervals of 5. Try graphing the results for a clearer understanding.
3. Collect some variable measures on items found around the house or at work such as the weight of individual raisins from a box, or the outside diameter of 1/4" washers. Use the data to redo Problems 1 and 2.
4. Generate some random numbers in Excel® or Minitab® and redo Problems 1 and 2.
5. In a normal distribution, what percentage of data fall within the area mentioned:
   a. Two standard deviations below the mean
   b. Between 1 standard deviation below the mean and 1 standard deviation above the mean
   c. Three standard deviations above the mean
   d. Below the mean
6. What is the z-score of a datum of 398 when the mean of the distribution is 455 and the standard deviation is 1.93?
7. Calculate and memorize the z-scores for the following probabilities under the z-table:
   a. .9
   b. .95
   c. .98

8. Give an example of each of the following:
    a. Positive skew
    b. Negative skew
    c. Platykurtic kurtosis
    d. Leptokurtic kurtosis

# FIVE

## SPC for Attribute Measures

From an operations perspective, the most common application of statistics in industry is by far Statistical Process Control (SPC). It was introduced by William Shewhart in 1924 at the Western Electric Company. SPC is known for its control charts that graph how a process stays on track in terms of natural variation, target values, reduction of inconsistencies, and ultimately quality. The process, however, refers to the methods of making a product — the material, machines, and a system of operations. SPC identifies common cause (natural) variation and special cause (assignable) variation. Problems in the process typically result in quality problems with the product, inefficiencies throughout the system, and inflated costs. With SPC, as the process begins to improve, it allows a better product quality, greater efficiency, and a reduced overall cost.

SPC was well thought out by World War II and helped in the war effort, but it did not fully make its way into industry until Edward Deming introduced Japanese manufacturers to SPC in the 1950s. Manufacturers then began to see the benefits of SPC. It took another thirty years for the rest of the world (including the United States) to take SPC seriously as a method of quality control.

SPC includes two distinct categories of charts: attribute type measurements and variable type measurements. SPC forms its statistical basis for attribute charts using discrete distributions (Chapter Three) and for variable charts using continuous distributions (Chapter Four). Both types of charts use the same method of implementation and both reveal problem areas in the process. In fact, with minor differences, both charts are generated using the same procedure.

This procedure generally requires that we define a particular product feature referred to in industry as a *quality characteristic*. After determining an appropriate measure of that quality characteristic, we then determine the number of data needed to generate the study as well as how the study will describe the process in terms of central tendency and dispersion. Finally, we plot the points on a graph, make improvements based on analyses of those graphs, and repeat the procedure to gain further improvement. Specific steps to create each type of charts and how to interpret them represent the majority of the content in Chapters Five and Six. But first, consider how to organize the data.

## Individual Data versus Subgroup Data

One distinct practice common to several charts in SPC is subgrouping data to normalize inconsistencies in the distribution. Sample distributions must share characteristics similar to the normal distribution before the normal distribution can be used to describe the area within those sample distributions. When a sample distribution is slightly skewed, has a slightly abnormal kurtosis, or displays multiple modes, it is less suitable to meet this condition; it must be forced into more of a normal shape.

This is accomplished by averaging a given consecutive set of data — called a *subgroup* — from the sample and utilizing that average as a single datum in a new distribution — called the *subgrouped distribution*. The subgrouped distribution displays a smaller distribution curve with a smaller standard deviation and range. With symmetrical distributions, the measures of central tendency remain equal to the initial distribution of individual values — subsequently called the *individual distribution*. With skewed, multimodal, and odd-shaped distributions, the subgrouped distribution becomes more normal in terms of dispersion. However, the subgrouped distribution must be used in conjunction with the individual distribution for any sort of further comprehensive analysis.

The individual distribution (those data not yet subgrouped) represents the product to a greater extent than the subgrouped distribution. Because the data in the subgrouped distribution are averaged from the individual distribution and have greater normality, the subgrouped distribution represents a better picture of the process. Because SPC seeks to improve the process, Quality Control (QC) charts are, therefore, based on the subgrouped distribution rather than the individual distribution.

Figure 5.1 shows both an individual distribution (n = 180) and a subgrouped distribution that uses a subgroup size of 5 (resulting in 36 values). Using raw, unsorted data, every five samples are averaged to create the subgrouped distribution. Note both distributions are close to normal based on simple observation. However, the individual distribution has a higher kurtosis than desired. Calculating skew and kurtosis shows the individual distribution has a moderately high excess kurtosis of 0.171 compared to a slightly high excess kurtosis of 0.078 in the subgrouped distribution.

Both show a slight skew, but are close enough to 0.0 to assume normality. However, the skew does shift to positive in the subgrouped distribution as opposed to the negative skew in the individual distribution. The calculated skew for the subgrouped distribution is 0.047 compared to the individual distribution of −0.035. Both distributions maintain an average of 50.03. Standard deviations are 10.05 for the individual distribution and 4.826 for the subgrouped distribution. This subgrouped distribution represents a distribution closer to normal; it is more suitable for calculating and analyzing probabilities related to the process.

Choosing a subgroup size is typically a simple matter in industry. However, the theory behind the difference a subgroup size makes is quite complicated and beyond the purpose of this text. For variable data, subgroup sizes of 4 or 5 are common. Attribute charts use substantially higher subgroup sizes. In either case, if the subgroup size is too large, the resulting distribution becomes too different than the individual distribution. It

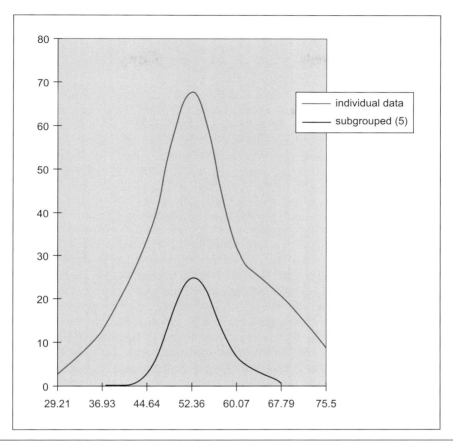

FIGURE 5.1  Individual distribution with n = 180 vs. distribution with subgroup size of 5

also may present a problem with too few points in the subgrouped distribution to show a legitimate curve. The subgrouped distribution should have at least 30 data points to support the measures of central tendency. Having a subgroup size too small does little to change the abnormalities in the individual distribution such as skew and multi modes.

Cost is an issue when considering subgroup size. Some tests can become quite expensive, consuming time and resources, and some tests are destructive or hazardous. Where costs are high, sample distributions are minimal, and subgroups will most likely be small.

## Common Control Charts for Attribute Measures

Attribute measures generate data that are typically 0 or 1, yes or no, go or no go, and so on. These were described in Chapter One as a distinct quality or characteristic of value. For example, attribute measures can describe if something is blue (1), or not blue (0). In industry, one common use is the go/no-go gauge. If the given characteristic fits, it is go; if it does not fit, it is no-go. As discussed in Chapter Three, the probability for a successful event with this type of measurement occurring is 0.5. Over multiple attempts with a continuous supply of product, measuring attribute data ultimately develops into a binomial distribution.

Two types of attribute charts are used in industry. The first type, the number of non-conforming units, counts the number of product units that are non-conforming. If the product unit has any kind of problem, the entire product unit is counted as non-conforming. The number of non-conforming units compared to the number of total units result in a proportion or simple count. This scenario results in $p$- and $np$-charts. The second type of attribute chart, count of non-conformities, counts how many non-conformities there are on each product unit. With this type, there could be more than one non-conformity on each product unit. The number of non-conformities is counted, not just the non-conforming unit. This scenario results in $c$- or $u$-charts.

## P-Chart

The most common control chart in industry using attribute data is the *p-chart*. The $p$ is the probability of the event occurring; it represents the proportion that is non-conforming within the subgroup. With the p-chart, as with any control chart, there are several steps to follow in creating the chart. They are:

1. Determine a quality characteristic to test.
2. Determine a rational subgroup.
3. Collect the data and plot the chart.
4. Compute the trial center line and control limits.
5. Remove unwanted variables and compute the revised center line and control limits.

With few differences, these same steps are followed not only with p-charts, but also with other charts covered later in this chapter and in Chapter Six. Consider these steps while moving through the following example.

A toy manufacturer produces whistles for children. This whistle is meant to be loud and easy to use with little effort. These features are regarded as quality characteristics from two aspects: children may not be able to produce a sustained, substantial flow of air, and children want a loud noise. These quality characteristics are actually the primary concern regarding the function of the product. In addition to the function, other quality characteristics include safety issues such as size (not too small to swallow), paint toxicity, price, durability, and ability to function under a variety of environmental conditions (wet, muddy, cold, etc.).

To test the primary quality characteristics (Step 1), the manufacturer designs an apparatus where the whistle is placed in position, blasted with air, and the loudness measured with a decibel meter. The air flow is adjusted to represent the average pressure produced by children of the target age, hence making the ease of use characteristic a constant. Now, the only quality characteristic tested is loudness. Through refinement of the design, the decibel meter is replaced with a decibel activated switch powering a light mounted on the console of the apparatus. This works because the whistle must only achieve a certain decibel level (the light) to meet the performance of the desired quality characteristic — the continuous meter complicated the testing procedure. The operator (wearing protective ear plugs) issues the blast of air with a button and checks for the light to illuminate. If the light illuminates, the whistle is good; if the light fails to illuminate, the whistle is bad. In practice, the terms *good* and *bad* are replaced with *conforming* and *non-conforming* respectively.

The manufacturer determines the kind of variation the test will expose as the whistles come off the production line; the source of that variation can then be controlled. This determination represents Step 2 in creating a quality control chart.

The first scenario is to gather a specified number of whistles all at once. This method exposes what is called a *piece-to-piece* variation from an isolated time frame. It exposes variation that may be occurring at a particular machine, with a particular machine operator, or during a particular shift or time of day. A second scenario is to gather a specified number of whistles one at a time, separated by a period of time. This method exposes what is called a *time-to-time* variation. It gives a good picture of the reliability of the process over a long period of time. A third scenario is called *within-piece* variation. This type of variation is seen within one product at a time; it is typically used in c- and u-charts or count of non-conformities charts Along with the assurance the sample is non-biased, random, and the subgroup size is statistically adequate, Step 2 determines what is referred to in industry as a *rational subgroup*.

When determining the subgroup size, the manufacturer relies on historical data regarding the probability of non-conforming data. If the subgroup size is too small, the chart will likely have several subgroups with zero non-conformity; the chart will then be meaningless. If the subgroup size is too large, they will face excess expense for data collection. For the whistle example, the manufacturer determines over an extended period of time that the proportion non-conforming from the current process is a probability of 0.05. Therefore, a subgroup size of 120 gives an average non-conformance of 6 whistles in each subgroup ($120 \times 0.05$), adequately representing any kind of variation that may be present in the process.

The whistle manufacturer decides to view the process quality from an overall perspective over a long period of time (time-to-time variation) and proceeds to Step 3 to create a quality control chart for collecting the data. Every four minutes, the operator collects one whistle randomly from a steady and continuous stream of product. The operator installs the whistle into the testing apparatus, tests the whistle, and records the time, date, and outcome (conforming or non-conforming with a probability of 0.5). The operator then notes any unusual observations during the test regarding the product or process. Actual output of the process is 460,000 whistles per year. The plant operates 270 days per year; each day consists of one shift of 480 minutes. At the end of one shift, the operator has collected one subgroup consisting of 120 out of an approximate total of 1704 whistles. During data collection, the operator installs and tests the product with care and consistent technique, treating each trial equally. This care is essential for preventing any unwanted variation from the data collection process from entering the study.

## Chapter Five

TABLE 5.1  Tabulated Data for Whistles

| Subgroup | Subgroup Size | np | p | Comments |
|---|---|---|---|---|
| 1 | 120 | 7 | 0.058 | |
| 2 | 120 | 4 | 0.033 | |
| 3 | 120 | 9 | 0.075 | |
| 4 | 120 | 10 | 0.083 | |
| 5 | 120 | 3 | 0.025 | |
| 6 | 120 | 6 | 0.05 | |
| 7 | 120 | 12 | 0.1 | |
| 8 | 120 | 5 | 0.042 | |
| 9 | 120 | 9 | 0.075 | |
| 10 | 120 | 9 | 0.075 | |
| 11 | 120 | 3 | 0.025 | |
| 12 | 120 | 8 | 0.067 | |
| 13 | 120 | 11 | 0.092 | Replaced Die (Worn) |
| 14 | 120 | 0 | 0 | |
| 15 | 120 | 5 | 0.042 | |
| 16 | 120 | 1 | 0.008 | |
| 17 | 120 | 0 | 0 | |
| 18 | 120 | 2 | 0.017 | |
| 19 | 120 | 8 | 0.067 | |
| 20 | 120 | 1 | 0.008 | |
| 21 | 120 | 4 | 0.033 | |
| 22 | 120 | 3 | 0.025 | |
| 23 | 120 | 8 | 0.067 | |
| 24 | 120 | 8 | 0.067 | |
| 25 | 120 | 4 | 0.033 | |
| 26 | 120 | 1 | 0.008 | |
| 27 | 120 | 12 | 0.1 | Hydraulic Pressure Incorrect |
| 28 | 120 | 2 | 0.017 | |
| 29 | 120 | 8 | 0.067 | |
| 30 | 120 | 7 | 0.058 | |
| 31 | 120 | 8 | 0.067 | |
| 32 | 120 | 10 | 0.083 | |
| 33 | 120 | 11 | 0.092 | |
| 34 | 120 | 8 | 0.067 | |
| 35 | 120 | 7 | 0.058 | |
| Σ | 4200 | 214 | 1.784 | |

To provide enough data for control, the operator collects data for the next 35 days (25 total subgroups is minimal). All data are tabulated as represented in Table 5.1 (see Appendix E).

Step 3 continues with creating a quality control graph, converting the chart (Table 5.1) into a meaningful, visual form (Figure 5.2). Each subgroup is plotted as one point on the chart. Note the stratification of the data along the *y*-axis between subgroups 9 and 10, and again between 23 and 24. This is caused by the discrete nature of the data. Subgroups with no non-conforming items result in a 0.0 proportion. One non-conforming unit within a subgroup results in a proportion of 0.008333, two 0.016667, three 0.025, and so on. The highest number non-conforming was 12, with a proportion of 0.1 found in subgroups 7 and 27. Only one subgroup (subgroup 6) had 6 non-conforming units, which fell at the center of the graph with a 0.05 proportion and was equal to the outgoing quality of the process. Note also the shape of the graph. The first 13 subgroups seem to be higher than the rest, and the remaining subgroups seem to trend upward.

Examining the data table is critical at this point to ascertain if there are any notes. These notes may help to explain why the chart is representing this uneven progression; they may uncover unwanted variation in the process. Determining where the unwanted variation is coming from, eliminating it, and retesting to show any improvement is at the heart of SPC quality control. The comments indicate the machine die is worn, causing an increase in non-conforming product. After the die is replaced (in time for subgroup 14) the quality level improves. The upward trend illustrates the continued use of the new die as it wears. Eventually this sequence will cycle over and over. Another note suggests a temporary operator that causes the machine to malfunction — the hydraulic pressure on one of the settings is incorrect (subgroup 27). This error causes an increase of non-conforming product until it is found.

Step 4 involves calculating temporary or trial limits of the chart, including the center line and the control limits. The center line is at or very near the established

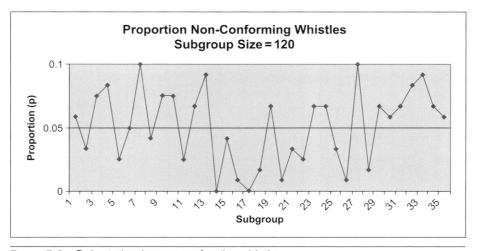

FIGURE 5.2  P-chart showing non-conforming whistles

outgoing quality level using the average of the subgroup proportions, hence establishing an overall probability. Using equation 2.1, these data have an average of:

$$\bar{x} = \Sigma p / N$$
$$= 1.784 / 35 = 0.050952$$

where:

$$N = \text{number of subgroups}$$

or:

$$\bar{x} = \Sigma np / Nn$$
$$= 214 / 4200 = 0.050952$$

where:

$$n = \text{number in each subgroup}$$

Either method will give the same answer. As expected, the center line is the same as the average quality level determined as 0.05 over a long period of time.

The control limits include the upper control limit (UCL) and the lower control limit (LCL). These limits are used to represent tails of a normal distribution at three standard deviations above the mean (Figure 5.3). If the subgroup exceeds the limit, the subgroup may have gone beyond a normal probability of existing on that point of the chart. Less than 2% of data in the normal distribution exist beyond three standard deviations from the mean. (This is only one of several conditions indicating an out-of-control situation. The other conditions are covered in detail in Chapter Seven.)

Therefore the control limits are to show three standard deviations above and below the established center line of the chart. The following equations are used to figure each of the limits respectively:

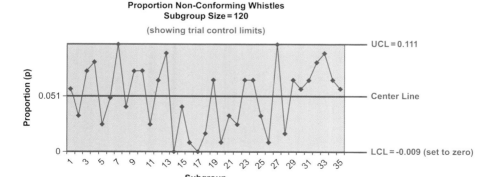

FIGURE 5.3  P-chart showing calculated trial values

## SPC for Attribute Measures

**Upper Control Limit**

$$\text{UCL} = \bar{p} + 3\sqrt{\bar{p}(1-\bar{p})/n} \qquad (5.1)$$

**Lower Control Limit**

$$\text{LCL} = \bar{p} - 3\sqrt{\bar{p}(1-\bar{p})/n} \qquad (5.2)$$

If three standard deviations below the mean return a negative number, the lower control limit is zero because there cannot be a negative LCL in attribute charts. Technically, subgroups existing on, around, or below the lower control limit in attribute charts is a good thing. It means there are few or no non-conforming items. Achieving zero defects in reality is impossible. However, if subgroups start to congregate toward the negative side of the chart, the operator should take note, determine the cause of the shift, and repeat whatever caused it — assuming the cause is not attributed to an error in data recording.

As for the whistle chart (Figure 5.3), note the calculated value of the LCL is:

$$\text{LCL} = \bar{p} - 3\sqrt{\bar{p}(1-\bar{p})/n}$$
$$= 0.050952 - 3\sqrt{0.050952(1-0.050952)/120} = -0.00927$$

and is set to 0.0. The UCL is:

$$\text{UCL} = \bar{p} + 3\sqrt{\bar{p}(1-\bar{p})/n}$$
$$= 0.050952 + 3\sqrt{0.050952(1-0.050952)/120} = 0.111174$$

There are no points exceeding these limits. But the limits established in this step are only trial or temporary limits. Because there is always room for improvement, any problems encountered during the study should be rectified before continuing, especially if they show a subgroup close to or on an upper control limit, such as subgroups 7 and 27.

There is no explanation for datum 7. The only way to correct this would be an investigation into what happened at the time this subgroup was collected. The two problems that are recorded reveal die wear and a faulty hydraulic setting on subgroups 13 and 27. These problems may or may not be fixed, depending on money. To fix the die problem, the company may need to spend considerable capital in using higher quality dies. Temporary worker problems may be solved through extended training, monitoring, or elimination of that source of labor. Each has its cost. For the sake of the example, however, management has decided all temporary workers must now undergo extra training, and new technology is installed to improve the wear of the dies. These two problems are solved and the study can continue.

Step 5, the final step, calculates the improved limits. It utilizes a simple process of eliminating the troubled subgroups from the formulas shown above. Fourteen subgroups are eliminated. Subgroups 1 through 13 are eliminated because of the worn die, and 27 because of the faulty hydraulic setting. The recalculated center line is:

## Chapter Five

$$\bar{p}_{adj} = (\Sigma np - \Sigma np_{nc}) / (\Sigma n - \Sigma n_{nc}) \tag{5.3}$$
$$= 106 / 2520 = 0.042063$$

where:

nc = non-conforming
adj = adjusted by eliminating the non-conforming subgroups

The recalculated control limits is:

$$\text{LCL} = \bar{p}_{adj} - 3\sqrt{\bar{p}_{adj}(1-\bar{p}_{adj}) / n} \tag{5.4}$$
$$= 0.042063 - 3\sqrt{0.042063(1-0.042063) / 120} = -0.01291$$

and:

$$\text{UCL} = \bar{p}_{adj} + 3\sqrt{\bar{p}_{adj}(1-\bar{p}_{adj}) / n} \tag{5.5}$$
$$= 0.042063 + 3\sqrt{0.042063(1-0.042063) / 120} = 0.097036$$

Note in Figure 5.4 how each limit is reduced because of the reduction in the center line; also, the overall span of limits is tightened. The LCL is set to 0.0 because there cannot be negative non-conformities. These newly calculated values become targets for the next segment of the study. The quality technician repeats the study using the same sampling methods as before. The new chart is graphed and compared to the first chart to visually examine any improvement made through the process and to expose further problems to eliminate.

The study should continue to be monitored in such a way until the quality level is satisfactory. Then a reduced level of monitoring is required. Over time, a chart comparison will reveal improvements from the beginning of the study (Figure 5.5).

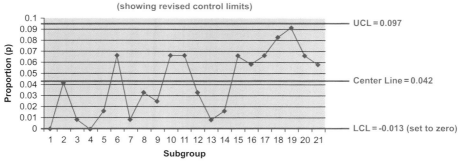

FIGURE 5.4   P-chart showing revised control limits

## SPC for Attribute Measures

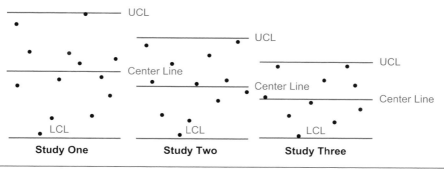

FIGURE 5.5   Chart improvements over time

Note the last 7 subgroups in Figure 5.4. They represent yet another out-of-control situation called a *run*. This type of condition and others will be discussed further in Chapter Seven.

### NP-Chart

*NP*-charts are similar to p-charts in that they count non-conforming product. The difference is the np column in Table 5.1 is graphed rather than the proportion column. Essentially this chart is identical to the *p*-chart, but may be easier for plant floor personnel to read and understand because there are no proportions or probabilities to calculate (Figure 5.6). However, *np*-charts present a mathematical problem when

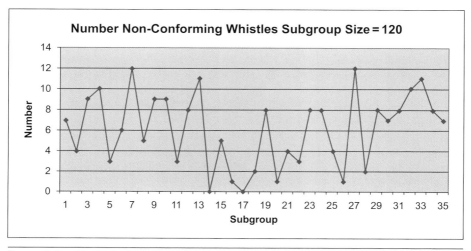

FIGURE 5.6   *NP*-chart of whistle data

subgroup size varies. Only *p*-charts can be used when counting non-conforming units if a variable subgroup size exists (see Chapter Seven).

*NP*-chart center lines and control limit calculations are slightly different than *p*-charts because proportions are replaced with actual numbers; however, their function remains the same. The following equations show the *np*-chart center line, the upper control limit, and the lower control limit for the whistle data.

$$n\bar{p} \tag{5.6}$$

$$n\bar{p} = 120/0.051 = 6.12$$

$$\text{UCL} = n\bar{p} + 3\sqrt{n\bar{p}(1-\bar{p})} \tag{5.7}$$

$$= 6.12 + 3\sqrt{6.12(1-0.051)}$$

$$= 6.12 + 3\sqrt{5.80788} = 6.12 + 5.409954356 = 11.53$$

$$\text{LCL} = n\bar{p} - 3\sqrt{n\bar{p}(1-\bar{p})} \tag{5.8}$$

$$= 6.12 - 3\sqrt{6.12(1-0.051)}$$

$$= 6.12 - 3\sqrt{5.80788} = 6.12 - 5.409954356 = 0.71$$

where,

$\bar{p}$ = average of the proportion column

## C-Chart

The c-chart is a count of non-conformities chart where one unit could have none, one, or more non-conformities. Note the non-conformities are all from one unit, meaning the subgroup size is one. This type of chart is used frequently on items such as cars, planes, and other large units. The procedure for creating a c-chart remains the same as for other control charts. The steps determine the quality characteristic, determine a rational subgroup (in the case of a *c*-chart, it is one), collect the data, plot the chart, compute the trial center line and control limits, remove unwanted variables, and compute the revised center line and control limits.

Consider a car manufacturer. As a final inspection, the car is scanned for any visual non-conformity. For example, the fender may not line up with the door, or the hood may have an uneven seam. The inspection may also include scratches, paint bubbles, loose rear-view mirror, seat blemishes, or any other noticeable non-conformity. After inspecting several cars in this manner, the manufacturer can reveal where in the process problems exist. Detailed data collection is of prime importance because of the variety of non-conformities. If certain problems begin to be common among cars, efforts are concentrated on that problem. If a subgroup is out of control or a set of subgroups display unusual patterns, and there is no single cause, an overall effort

# SPC for Attribute Measures

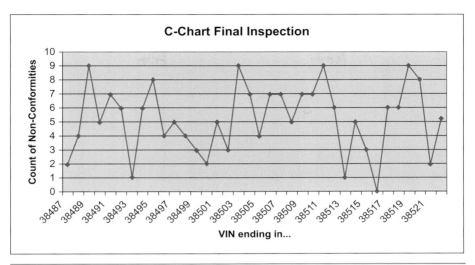

FIGURE 5.7   C-chart of cars

to increase quality is warranted. Figure 5.7 shows the result of collecting 182 non-conformities on 35 vehicles. Note the similarity to the other attribute charts. Also note that the subgroup consists of one unit labeled by the car's VIN. In this case, the c-chart subgroup denotes one car.

Subgroups can frequently denote a period of time such as a day, week, or month. C-charts are commonly applied to situations where a count of non-conformities is tallied through that time frame. An example is an automobile assembly plant counting paint scratches in the 3rd week of June. When multiple non-conformities are present, some non-conformities may be more serious than others. For example, scratches may be considered less serious than paint bubbles. Therefore, scratches might receive a weight of 1 whereas bubbles would be weighted 2, and misaligned fenders or hoods weighted 3.

Calculations for both the trial and revised (non-weighted) chart values follow. The center line is:

$$\bar{c} = \Sigma c / N \tag{5.9}$$

where

$N$ = the total number of subjects in the study (35 cars)

For the car example:

$$\bar{c} = 182 / 35 = 5.2$$

The upper and lower control limits are:

$$UCL = \bar{c} + 3\sqrt{\bar{c}} \tag{5.10}$$

and:
$$LCL = \bar{c} - 3\sqrt{\bar{c}} \qquad (5.11)$$

For the car example:
$$UCL = 5.2 + 3(2.280351) = 12.04105$$

and:
$$LCL = 5.2 - 3(2.280351) = -1.64105 \text{ (set to 0.0)}$$

Revised values recalculated for corrections are similar to other charts. This study reveals no out-of-control points or patterns. If there are, those points revealing a problem will be removed from the equations after a solution is reached. In the case of the equations above, the only one that will change is the center line (equation 5.9). Once the new center line is established, it will be used as usual in the upper and lower control limit equations.

If the four highest data are eliminated from the car example above, the recalculated values become:
$$\bar{c} = 146/31 = 4.709677$$
$$UCL = 4.709677 + 3(2.170179) = 11.22021$$

and:
$$LCL = 4.709677 - 3(2.170179) = -1.80086 \text{ (set to 0.0)}$$

## *U*-Chart

Similar to the *c*-chart, the *u*-chart is a count of non-conformities chart. However, the *u*-chart has more than one unit per subgroup and looks at non-conformities per unit (hence the name "u" chart). This approach accommodates a count of non-conformities chart with the possibility of a varying subgroup size. Of the four attribute charts mentioned so far, only the *p*- and the *u*-charts can accommodate variable subgroup sizes. Variable subgroup size is a topic for Chapter Seven.

When the subgroup size is uniform, the construction and function of a *u*-chart is identical to the *c*-chart, other than the calculation of *u*. Calculate *u* by dividing the number of subjects in the subgroup by the number of non-conformities ($c$). For example, if there are 100 units in a single subgroup, these units are checked for several possible non-conformities each, and the result is 110 non-conformities for the entire subgroup, u is:
$$u = c/n = 110/100 = 1.1$$

The u value is determined for every subgroup and plotted on the chart. The chart value equations are almost identical to the *c*-chart. Consider the differences:
$$\bar{u} = \Sigma c / \Sigma n \qquad (5.12)$$

$$UCL = \bar{u} + 3\sqrt{\bar{u}/n} \qquad (5.13)$$

and:

$$LCL = \bar{u} - 3\sqrt{\bar{u}/n} \qquad (5.14)$$

## Chapter Five Discussion

1. Discuss the historical background of SPC? Who are the main people responsible for this development? When did these developments occur?
2. Explain why Japan developed much faster than the United States regarding quality.
3. What is the main purpose of SPC and what are the steps to implement SPC?
4. Explore the differences between individual data and subgrouped data.
5. List and explain the differences and uses of the most common types of attribute charts.
6. Explain the differences among piece-to-piece, within piece, and time-to-time variation. Elaborate on when these would be used.
7. Explain number non-conforming versus count of non-conformities.
8. When are each of the four attribute charts (discussed in this chapter) used in industry?

## Chapter Five Problems

1. Collect several data that are easy to obtain (about 120), tabulate them, calculate the descriptive statistics, and graph them into a histogram. Then separate the data into groups of 4 or 5 and redo the statistics and histogram. Explain the difference.
2. Use the Excel® and Minitab® data set provided for this problem (see Appendix E). Subgroup into groups of 20, tabulate, calculate proportions, and construct a $p$-chart and an $np$-chart.
3. Use the data in Problem 2 to represent the count of non-conformities. Create a $c$-chart and a $u$-chart to see the results. (Hint: For the $c$-chart do not subgroup the data; use each number as one unit).
4. Find something around the house that is in abundance (bag of bean, box of nails, toothpicks) and go through the steps to create an attribute SPC chart to reveal the quality of the items. Once collected, tabulate, calculate the statistics, and enter the data into a $p$-chart and/or an $np$-chart.
5. Repeat the initial process from Problem 4. However, this time look for more than one attribute per item. Then collect, tabulate, calculate the statistics, and enter the data into a $u$-chart and/or $c$-chart.

# SIX
## SPC for Continuous Measures

Discrete measures such as people, cars, planes, etc. must be viewed in whole, as 1, 2, 3, and so on. One cannot, for example, have 2.5 children. Continuous measures on the other hand may be viewed in part; that can be carried out to a decimal accuracy. Length, weight, volts, and volume are some of the most obvious examples. The average height of an adult male in the United States is 5′ 9.2″. Because height is a variable and can include infinite positions between the nominal measure, an adult U.S. male can actually be 5′ 9.2″. With a more accurate means of measuring height, along with strict criteria for collecting data, he could be 5′ 9.274682″. Heights are not restricted to discrete measures such as 5′ 9″ or 5′ 10″.

Weight is similar. People don't jump from 179 lbs. to 180 lbs. They increase continuously, although perhaps rapidly, between these two values while hitting an infinite number of decimal places along the way. In other words, it is possible to be 179.988764 lbs. The significance of the decimal place is another issue. Although someone can be 179.988764 lbs., rarely would they report that other than 180 lbs. The 0.011236 difference is considered insignificant — when measuring a person's weight. However, in cases where a millionth of a pound *is* significant, and assuming there is the ability to measure to this accuracy, values would not be rounded to the next highest number.

Consider a cylinder head for a four-cylinder internal combustion engine. If the head surface is measured with a handheld ruler, the reading may give an accuracy to the 16th or 32nd of an inch. However, this reading reveals nothing about the quality of the head. The ability of the head to operate properly is contingent on accuracies in the thousandths — distances not discernible to the unaided human eye. To measure to this level, the machinist uses feeler gauges, a micrometer, or an electronic means of measuring this distance.

In a manufacturing setting for casting, surfacing, and preparing cylinder heads for car motors, this measure is critical for ascertaining the quality of the product. But manufacturing distances to this kind of accuracy requires a delicate process and individual items vary from one to the other. They are much like fingerprints. Although close, there are never two items that are identical. Over a period of time, a natural pattern emerges when manufacturing these heads where this distance varies between limits.

This is the continuous distribution introduced in Chapter Four. Quality technicians use this distribution and subgrouped distributions to explain the product, uncover problems with the process, and improve quality.

The three most common variable charts used in industry are the X-Bar ($\bar{x}$) chart, the R chart, and the S chart. ($\bar{x}$ typically denotes the mean.) The steps in creating these charts are similar to the attribute charts discussed in Chapter Five. They include determining the quality characteristic, determining a rational subgroup, collecting the data, plotting the chart, computing the trial center line and control limits, removing unwanted variables, and computing the revised center line and control limits. The primary difference between variable charts and attribute charts are the techniques in collecting data; fewer subgroups are typically required.

## Control Charts for Variable Data

### X-Bar and R Charts

As with any SPC study, the X-Bar chart begins with a quality characteristic. This first question determines what kind of data is present (attribute or variable). If variable, what methods of measurement are available for the required accuracy? A measure of the height of wooden fence posts would require only a tape measure. Trying to measure distances varying between thousandths of an inch with a tape measure, such as cylinder head flatness, would yield one similar value and show no variance. Only after this initial consideration is made and the measuring instrument is determined, the study can proceed to the next step.

Consider once again the example of the cylinder heads. The cylinder head is a major component of any internal combustion engine — the kind of engine in a car. It is bolted to the top of the cylinder block, where the pistons are. The head is responsible for maintaining compression on the top end of the motor (Figure 6.1). The variation along the surface of the head that comes in contact with the block of the motor must be within specifications for the motor to perform adequately. If the variation is too great, the motor will lose compression, will overheat, and will eventually stop running altogether. If the variation is too little, the cost associated with manufacturing the head to such high tolerances is too high and profits, if not demand for the product, disappear.

The quality of this surface is referred to as flatness; it can be measured with feeler gauges underneath a straight edge positioned in various locations of the surface. The feeler gauge gives a distance accurate to the thousandths measuring any displacement between the straightedge and the head surface. A much more accurate and precise means of measuring this flatness is to use a laser displacement gauge with accuracy beyond the 1/100,000th of an inch. For the sake of simplifying the example, this study will consider measurements along one line only, changing the quality characteristic from flatness to straightness. This characteristic still represents a distance. Distances are variables. So, the quality characteristic of the surface of a head requires a variable study utilizing variable charts with distance measures.

SPC for Continuous Measures  **105**

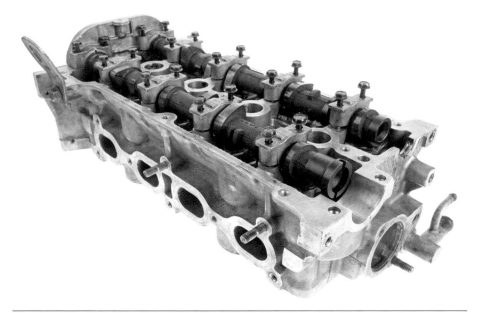

FIGURE 6.1   Internal combustion engine cylinder head

Measuring each individual head coming off the production line is too expensive and not necessary for the study. As discussed in Chapter Five, a subgroup size of 4 is fine. Each subgroup is collected at one time by selecting four consecutive heads from the milling process. A period of time elapses and the next collection occurs. This method results in a piece-to-piece variation that minimizes the variation between heads in the subgroup while maximizing the variation between subgroups. This type of rational subgroup best represents how the process of surfacing heads varies.

The study initially consists of 40 subgroups where measurements are the maximum clearance through a centerline on the long side of the head. Experienced technicians place the head into the measuring position and calibrate the instrument. The laser essentially maps the critical surface of the head, and the displacement between a theoretical flat surface and the surface of the head results in a distance measure for each datum. These data are recorded into each subgroup with the date, time, and any comments that may help to uncover unnatural variance existing in the study. Typically, the technician records the data, documents any pertinent events with comments, computes averages ($\bar{x}$) and ranges (R) of each subgroup, and plots all points on one worksheet.

**X-Bar Chart**   In this example, Figure 6.2 shows a typical data recording worksheet (in this case subgroups 1 through 12 only), Figure 6.3 shows the trial X-Bar chart, and Figure 6.4 shows the corrected X-Bar chart. These worksheets vary in design, but most give areas for computations and show standard constants regarding limits. They are available from several sources including the American Society of Quality (ASQ).

## 106 Chapter Six

**Xbar and R Chart for Cylinder Head Data:**
Operator: D. Jackson
Date: 10/01/2014
Time: 7:46am          Data: (4 units measured every 10 minutes)

| Subgroup | x1 | x2 | x3 | x4 | Xbar | Range | Comments |
|---|---|---|---|---|---|---|---|
| 1 | 0.000419 | 0.00198 | 0.001117 | 0.001412 | 0.001232 | 0.001561 | machine warm up |
| 2 | 0.000799 | 0.001336 | 0.001245 | 0.001866 | 0.001312 | 0.001067 | " |
| 3 | 0.00104 | 0.001741 | 0.00111 | 0.001015 | 0.001227 | 0.000727 | " |
| 4 | 0.001749 | 0.001463 | 0.000998 | 0.000673 | 0.001221 | 0.001076 | |
| 5 | 0.001401 | 0.001023 | 0.000947 | 0.00158 | 0.001238 | 0.000633 | |
| 6 | 0.00138 | 0.001123 | 0.000445 | 0.001535 | 0.001121 | 0.00109 | |
| 7 | 0.001486 | 0.001289 | 0.001081 | 0.001575 | 0.001358 | 0.000494 | |
| 8 | 0.000204 | 0.001053 | 0.000885 | 0.001627 | 0.000942 | 0.001424 | bumped laser |
| 9 | 0.000518 | 0.000581 | 0.000268 | 0.000594 | 0.00049 | 0.000327 | |
| 10 | 0.000486 | 0.001331 | 0.001118 | 0.00106 | 0.000999 | 0.000846 | |
| 11 | 0.001748 | 0.000939 | 0.00176 | 0.00137 | 0.001454 | 0.000822 | |
| 12 | 0.00081 | 0.000689 | 0.001037 | 0.001055 | 0.000898 | 0.000366 | |

FIGURE 6.2    Sample X-Bar R chart worksheet

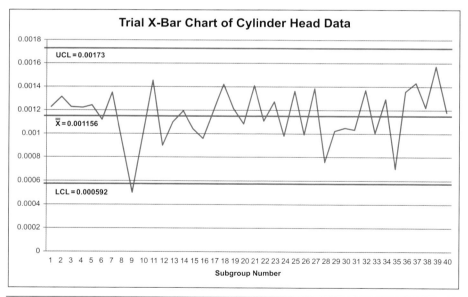

FIGURE 6.3    X-Bar chart with trial center line and limits

Software is available too in abundance. The quality technicians should make a survey of what is available and choose what best serves their purpose. Resourceful technicians could even make their own worksheet using a standard spreadsheet such as Excel® or statistics-oriented software such as Minitab®.

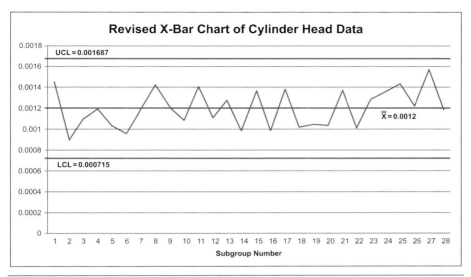

FIGURE 6.4  X-Bar chart with revised center line and limits

The next step in this process involves the calculations, first of which is the trial center line for the X-Bar chart. As it was with the other control charts, the centerline is the average of the subgroups. Because the subgroups are averages themselves, the centerline of an X-Bar chart is an average of the averages. However, the centerline is virtually equal to the average of the individual distribution. The centerline is computed with:

$$\bar{x} = \Sigma x/n \qquad (2.1)$$

where x represents each subgroup, and n is the number of subgroups. Technically, X-Bar ($\bar{X}$) becomes X-Double Bar ($\bar{\bar{X}}$) because this value represents the average of the averages; on occasion, it is labeled as such. Using the data from the example, the centerline becomes:

$$\bar{\bar{X}} = 0.04622/40 = 0.001156$$

The second set of calculations is for the trial upper and lower control limits. The limits are placed at three standard deviations (of the subgroup distribution) above and below the centerline respectively. The upper and lower control limits (trial, revised, or any subsequent values) theoretically represent $3\sigma$ in the distribution. In simple terms, this suggests $CL = \pm 3\sigma$. It is unlikely there is knowledge of $\sigma$ (or $\mu$ for that matter); therefore, standard chart constants found in Figure 6.4 are used as estimates. The trial control limits are calculated as follows:

$$UCL = \bar{\bar{X}} + A_2 \times \bar{R} \tag{6.1}$$

and

$$LCL = \bar{\bar{X}} - A_2 \times \bar{R} \tag{6.2}$$

where X-Double Bar is the mean of all the subgroups and R-Bar is the mean of the subgroup ranges. For these data, the mean of the subgroup ranges is 0.000781.

$A_2$ and the other values in the table shown in Figure 6.5 are based on subgroup size. In this case, the subgroup size is 4 (found in the first column, 3rd row of values), so the value for $A_2$ is 0.729. These constants are used extensively for many statistical chart applications in industry. Their derivation goes beyond the scope of this text. But an enlightening exercise for the student is to calculate the values for the cylinder head example using these constant values and then compare them to the results given using $CL = \pm 3\sigma$. There is a slight difference in the results, especially for small subgroup sizes. The difference these constants account for is a closer estimation of sigma in a process sometimes referred to as *unbiasing* the sample. Using these constants can easily be incorporated into any number of software packages used in SPC.

Following through with the example, the upper and lower trial limits are:

$$UCL = \bar{\bar{X}} + A_2 \times \bar{R} =$$
$$= 0.001156 + 0.729(0.000781) = 0.00173$$

and

$$LCL = \bar{\bar{X}} - A_2 \times \bar{R}$$
$$= 0.001156 + 0.729(0.000781) = 0.000592$$

Once the trial center line and limits are established and inserted into the chart, study the chart carefully and determine where areas of interest are. In this example (Figure 6.3), two areas of attention arise during the first 9 or 10 subgroups; the data worksheet can point to some of the answers. The first area is explained by a system warm up. Probably the best course of action is to wait until the machine is fully functional before measuring parts. Without the comment in the worksheet, this condition would have given the erroneous impression that parts were particularly close to the average in these subgroups. The second questionable area is with subgroup 8 where the measuring table was bumped. Again this explanation comes from a comment on the worksheet. Apparently, this condition was not fully corrected and most likely resulted in the out-of-control point representing subgroup 9.

Later in the chart there are some interesting fluctuations, but still within the control limits. With no comments regarding this section, the quality technician can investigate to find a cause. But, it is probable a cause may not be found. The most desirable section of the chart is in the middle between subgroups 14 and 27. Something — not commented on in the notes — caused less variability and a general trend toward the centerline. This, of course, is desirable. The quality technician should again investigate and try to determine why. If successful in determining the cause, the process should be adjusted to reflect this improvement permanently if cost can be maintained.

CONTROL CHART CONSTANTS for SPC

| n | A | $A_2$ | $A_3$ | $c_4$ | $B_3$ | $B_4$ | $B_5$ | $B_6$ | $d_2$ | $1/d_2$ | $d_3$ | $D_1$ | $D_2$ | $D_3$ | $D_4$ |
|---|---|---|---|---|---|---|---|---|---|---|---|---|---|---|---|
| 2 | 2.121 | 1.88 | 2.659 | 0.7979 | 0 | 3.267 | 0 | 2.606 | 1.128 | 0.8862 | 0.853 | 0 | 3.686 | 0 | 3.267 |
| 3 | 1.732 | 1.023 | 1.954 | 0.8862 | 0 | 2.568 | 0 | 2.276 | 1.693 | 0.5908 | 0.888 | 0 | 4.358 | 0 | 2.575 |
| 4 | 1.5 | 0.729 | 1.628 | 0.9213 | 0 | 2.266 | 0 | 2.088 | 2.059 | 0.4857 | 0.88 | 0 | 4.698 | 0 | 2.282 |
| 5 | 1.342 | 0.577 | 1.427 | 0.94 | 0 | 2.089 | 0 | 1.964 | 2.326 | 0.4299 | 0.864 | 0 | 4.918 | 0 | 2.114 |
| 6 | 1.225 | 0.483 | 1.287 | 0.9515 | 0.03 | 1.97 | 0.029 | 1.874 | 2.534 | 0.3946 | 0.848 | 0 | 5.079 | 0 | 2.004 |
| 7 | 1.134 | 0.419 | 1.182 | 0.9594 | 0.118 | 1.882 | 0.113 | 1.806 | 2.704 | 0.3698 | 0.833 | 0.205 | 5.204 | 0.076 | 1.924 |
| 8 | 1.061 | 0.373 | 1.099 | 0.965 | 0.185 | 1.815 | 0.179 | 1.751 | 2.847 | 0.3512 | 0.82 | 0.388 | 5.307 | 0.136 | 1.864 |
| 9 | 1 | 0.337 | 1.032 | 0.9693 | 0.239 | 1.761 | 0.232 | 1.707 | 2.97 | 0.3367 | 0.808 | 0.547 | 5.394 | 0.184 | 1.816 |
| 10 | 0.949 | 0.308 | 0.975 | 0.9727 | 0.284 | 1.716 | 0.276 | 1.669 | 3.078 | 0.3249 | 0.797 | 0.686 | 5.469 | 0.223 | 1.777 |
| 11 | 0.905 | 0.285 | 0.927 | 0.9754 | 0.321 | 1.679 | 0.313 | 1.637 | 3.173 | 0.3152 | 0.787 | 0.811 | 5.535 | 0.256 | 1.744 |
| 12 | 0.866 | 0.266 | 0.886 | 0.9776 | 0.354 | 1.646 | 0.346 | 1.61 | 3.258 | 0.3069 | 0.778 | 0.923 | 5.594 | 0.283 | 1.717 |
| 13 | 0.832 | 0.249 | 0.85 | 0.9794 | 0.382 | 1.618 | 0.374 | 1.585 | 3.336 | 0.2998 | 0.77 | 1.025 | 5.647 | 0.307 | 1.693 |
| 14 | 0.802 | 0.235 | 0.817 | 0.981 | 0.406 | 1.594 | 0.399 | 1.563 | 3.407 | 0.2935 | 0.763 | 1.118 | 5.696 | 0.328 | 1.672 |
| 15 | 0.775 | 0.223 | 0.789 | 0.9823 | 0.428 | 1.572 | 0.421 | 1.544 | 3.472 | 0.288 | 0.756 | 1.203 | 5.74 | 0.347 | 1.653 |
| 16 | 0.75 | 0.212 | 0.763 | 0.9835 | 0.448 | 1.552 | 0.44 | 1.526 | 3.532 | 0.2831 | 0.75 | 1.282 | 5.782 | 0.363 | 1.637 |
| 17 | 0.728 | 0.203 | 0.739 | 0.9845 | 0.466 | 1.534 | 0.458 | 1.511 | 3.588 | 0.2787 | 0.744 | 1.356 | 5.82 | 0.378 | 1.622 |
| 18 | 0.707 | 0.194 | 0.718 | 0.9854 | 0.482 | 1.518 | 0.475 | 1.496 | 3.64 | 0.2747 | 0.739 | 1.424 | 5.856 | 0.391 | 1.609 |
| 19 | 0.688 | 0.187 | 0.698 | 0.9862 | 0.497 | 1.503 | 0.49 | 1.483 | 3.689 | 0.2711 | 0.733 | 1.489 | 5.889 | 0.404 | 1.596 |
| 20 | 0.671 | 0.18 | 0.68 | 0.9869 | 0.51 | 1.49 | 0.504 | 1.47 | 3.735 | 0.2677 | 0.729 | 1.549 | 5.921 | 0.415 | 1.585 |
| 21 | 0.655 | 0.173 | 0.663 | 0.9876 | 0.523 | 1.477 | 0.516 | 1.459 | 3.778 | 0.2647 | 0.724 | 1.606 | 5.951 | 0.425 | 1.575 |
| 22 | 0.64 | 0.167 | 0.647 | 0.9882 | 0.534 | 1.466 | 0.528 | 1.448 | 3.819 | 0.2618 | 0.72 | 1.66 | 5.979 | 0.435 | 1.565 |
| 23 | 0.626 | 0.162 | 0.633 | 0.9887 | 0.545 | 1.455 | 0.539 | 1.438 | 3.858 | 0.2592 | 0.716 | 1.711 | 6.006 | 0.443 | 1.557 |
| 24 | 0.612 | 0.157 | 0.619 | 0.9892 | 0.555 | 1.445 | 0.549 | 1.429 | 3.895 | 0.2567 | 0.712 | 1.759 | 6.032 | 0.452 | 1.548 |
| 25 | 0.6 | 0.153 | 0.606 | 0.9896 | 0.565 | 1.435 | 0.559 | 1.42 | 3.931 | 0.2544 | 0.708 | 1.805 | 6.056 | 0.459 | 1.541 |

FIGURE 6.5  Standard constants used in computing SPC charts
(Reprinted, with permission, from ASTM STP15D-Manual on Presentation of Data and Control Chart Analysis, copyright ASTM International, 100 Barr Harbor Drive, West Conshohocken, PA 19428)

## 110  Chapter Six

To update the chart to reflect the improvements, simply remove those subgroups. By eliminating subgroups 1 through 10, 28, and 35 (after problem determination and process improvement), a new chart may duplicate the desired area in the chart. However, this new chart may include some of the discarded units stemming from the measurement errors if they were re-measured. Because the error came from the measuring instrument and not the process, re-measuring those data represents a viable alternative to discarding those particular problem subgroups. Those subgroups where the error originated with the process have to be eliminated and the process must be improved prior to measuring further units.

Suppose the problem subgroups are eliminated. In the next step, new calculations of the centerline and control limits will expose variation closer to natural variance in the process (Figure 6.5) as opposed to assignable variation. The premise is that the true quality of the process is hidden by the errors occurring in the process or, as in this case, both the process and the measuring instrumentation. Eliminating these errors reveals the process's true quality. As the true quality of the process is revealed, improvements are made. This procedure continues as long as it leads to an improvement in quality.

The centerline of the revised chart is calculated with the same equation as before. The revised values of the two control limits require new equations found below. Again, using equation 2.1, the centerline for the revised chart becomes:

$$\bar{\bar{X}} = 0.033622/28 = 0.001201$$

This is not only the centerline, but also the X-Double Bar. Because it is the mean of the subgroup means, it is also called the *grand mean*. The R-Bar (mean of ranges) must be calculated as well after removing the aforementioned subgroups from the study. In this case, the new mean of the subgroup ranges is 0.000667. The formulae for the control limits using the chart constants are:

$$UCL = \bar{\bar{X}}_o + A\sigma_o \qquad (6.3)$$

and

$$LCL = \bar{\bar{X}}_o - A\sigma_o \qquad (6.4)$$

where $\bar{\bar{X}}_o$ is the new grand mean of the subgroups, $\sigma_o$ is an estimated sigma using R-Bar above (0.000667), and $A$ is 1.5 from Figure 6.5. First, $\sigma_o$ becomes:

$$\sigma_o = R_o/d_2 \qquad (6.5)$$

$$= 0.000667/2.059 = 0.00032417$$

where $R_o$ is the new R-Bar (mean of the ranges), and $d_2$ is 2.059 from Figure 6.5.

$$UCL = \bar{\bar{X}}_o + A\sigma_o$$

$$= 0.001201 + 1.5(0.00032417) = 0.001687$$

and
$$LCL = \bar{\bar{X}}_o - A\sigma_o$$
$$= 0.001201 - 1.5(0.00032417) = 0.000715$$

Note the tightening of the limits and the shift in the centerline (Figure 6.4). This movement in the limits and centerline represents the refinement of both the instrumentation and the process. It is important to note not only how the process can benefit by uncovering unwanted variance using this method, but also how measuring problems can be revealed and should typically be a part of any study. Repeating this study will likely reveal further improvements in the process. As a continuous action, this series of studies will continue requiring critical analysis of the process each time a chart is created, representing a continuous process improvement. To refine the measurement procedures and instrumentation, the study may include what is called a *Gage R&R* to control the reliability and repeatability of the measuring instruments. Computing a Gage R&R for this example goes beyond the scope of this text.

X-Bar charts are not complete without their companion R charts. As the X-Bar data reveal the measures of central tendency in the distribution, the R data reveal a measure of dispersion. One without the other shows only half the information regarding the distribution. As such, these two charts are always seen together.

**R Chart**   The R chart generated by the same cylinder head data in this example must be analyzed in conjunction with the X-Bar chart. Each subgroup is a distribution. As stated in Chapter Four, understanding any continuous distribution requires a value of central tendency (average) and a value of dispersion — in this case, the range. But the logic behind the R chart is slightly different than its X-Bar counterpart. In the X-Bar chart, the desired observation is close to the centerline. Greater distance from the centerline in either direction indicates variance. Subgroup values close to or over the limits indicate an unnatural source (assignable cause) of variance. Although R charts have centerlines, quality technicians view them as an expected result. The most desirable value with an R chart is close to or on the lower control limit. Quite frequently, this lower control limit is zero. When subgroup values approach this limit, there is little difference in measured values within the subgroup. In other words, product is coming out of the process with some definitive consistency. With an ability to control consistency — given an increase in quality — manufacturers can deliver high quality consistently.

Consistency, however, does not mean it is in the desirable location in regard to the centerline on the X-Bar chart. If the high quality part of the equation is missing, these items can be consistently of poor quality. For example, a subgroup may have a very low range, but actually be out of control in the X-Bar chart. This would simply mean product measures in this subgroup were very consistently out of control.

Conversely, a subgroup right on the centerline in the X-Bar chart may have a very large range and be out of control in the R chart. Imagine measurements in a subgroup (n = 4) that vary significantly, as in subgroup 30 in the R chart portion of Figure 6.6. In examining the data, it is revealed two of the values are not too far from the average ($\bar{R}$), but the other two include one very high and another very low. The X-Bar chart shows the subgroup as close to average (near the centerline). This problem subgroup — and out-of-control condition — can only be revealed by looking at the R chart and seeing the wide dispersion within this subgroup.

An ideal observation, with a combined examination of X-Bar and R charts, is one on the centerline in the X-Bar chart and close to or at the lower control limit in the R chart. The closest subgroup to this description in Figure 6.6 is subgroup 40. Quality personnel must determine any deviations from this level beyond what can be considered natural variation in the study.

Trial centerline calculation is similar to that for the X-Bar chart. The control limits are to be calculated using the constant values in Figure 6.5. To continue with the cylinder head example, the center line for the R chart is:

$$\bar{R} = \Sigma r/n \qquad (6.6)$$
$$= 0.031244/40 = 0.000781$$

and the limits are:

$$UCL = D_4 \bar{R} \qquad (6.7)$$
$$= 2.282(0.000781) = 0.001782$$

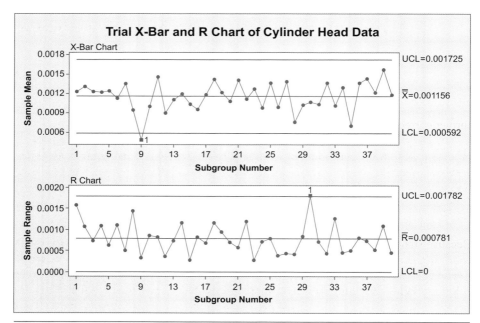

FIGURE 6.6  Trial X-Bar and R chart of cylinder head data

and
$$LCL = D_3 \bar{R} \tag{6.8}$$
$$= 0(0.000781) = 0$$

where $D_4$ and $D_3$ are found in Figure 6.5 for a subgroup of 4.

Upon examination of the R Chart, the prudent technician can see there may be a problem with subgroup 30. With this information, some investigation can occur with a plan of quality improvement. Recall during the X-Bar Chart revision that subgroups 1 through 10, 28, and 35 were removed. Removing subgroup 30 adds a further refinement to the process. In addition, note the changes to the X-Bar Chart after removing subgroup 30. Using equation 6.6, the corrected centerline is:

$$\bar{R} = \Sigma r/n$$
$$= 0.032575/27 = 0.000713$$

The equations for the revised R Chart again refer to constant values from Figure 6.5 and are:

$$UCL = D_2 \sigma_o \tag{6.9}$$
$$= 4.698(0.000346) = 0.001626$$

and
$$LCL = D_1 \sigma_o \tag{6.10}$$
$$= 0(0.000346) = 0$$

where $\sigma_o$ is the estimated sigma using R Bar (0.000713), and is computed using equation 6.5 as:

$$\sigma_o = R_o/d_2$$
$$= 0.000713/2.059 = 0.000346$$

The revised chart is shown in Figure 6.7 together with the X-Bar chart. Note the further corrections to the X-Bar Chart.

These procedures will continue by examining the charts, making corrections to the process based on the charts and observations of the process, revising the charts once assigned variance is revealed, and noting the improvement. Over time, improvements will be noticeable with an examination of historical charts.

## Standard Deviation Charts (S charts)

Because standard deviation is a measure of dispersion, the S chart can replace the R chart. In many cases, the S chart is more descriptive than the R chart, and more sensitive to changes in the dispersion of the distribution. S charts typically require a larger subgroup size — anything with 10 or more units within a subgroup. Traditionally they required laborious calculations, but computers often make it feasible to make a transition from the more common R chart to the more descriptive S chart.

## 114 Chapter Six

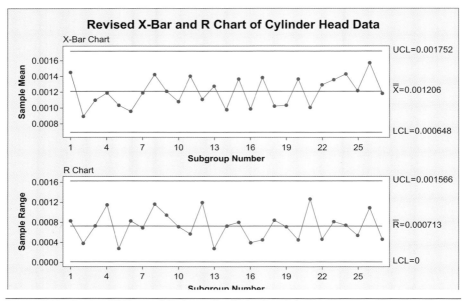

FIGURE 6.7  Revised X-Bar and R chart of cylinder head data

Because the S chart is a measure of dispersion, it is used in conjunction with the X-Bar chart, which measures central tendency. Almost everything about the X-Bar portion of the X-Bar and S chart is the same as with the X-Bar and R chart. However, the standard deviation is calculated for the dispersion measure in the distribution instead of the range. As data are collected, equation 5.6 is used to calculate s.

For charting purposes, the average of s becomes the centerline; in theory, the control limits become the centerline plus or minus 3 standard deviations of the s data. Otherwise, the procedure for creating the S chart is the same as it is for the other charts: Determine quality characteristic, select a rational subgroup, create the trial limits, and analyze the chart to remove any problem subgroups. Then revise the chart based on new calculations without those subgroups. The simplest method for calculating these values is to use the standard constants from Figure 6.5; here, the subgroup is at 10.

Through the efforts of a quality improvement team, the decision is made to study the cylinder head process in more detail by collecting more data and creating an S chart to show greater sensitivity from the process. Figure 6.8 shows the X-Bar and S charts resulting from this new study. There are still 40 subgroups in all, but with 10 measures in each subgroup. Referring again to Figure 6.5, the trial calculations for both the X-Bar and S charts are:

$$\bar{\bar{X}} = \Sigma \bar{x}/n \qquad (2.1)$$
$$= 0.454548/40 = 0.001136$$

$$\text{UCL}_x = \bar{\bar{X}} + A_3\bar{s} \qquad (6.11)$$
$$= 0.001136 + 0.975(0.000519) = 0.001643$$

$$\text{LCL}_x = \bar{\bar{X}} - A_3\bar{s} \qquad (6.12)$$
$$= 0.001136 - 0.975(0.000519) = 0.00063$$

$$\bar{s} = \Sigma s/n \qquad (6.13)$$
$$= 0.02076/40 = 0.000519$$

$$\text{UCL}_s = B_4\bar{s} \qquad (6.14)$$
$$= 1.716(0.000519) = 0.0008908$$

and

$$\text{LCL}_s = B_3\bar{s} \qquad (6.15)$$
$$= 0.284(0.000519) = 0.0001473$$

After the quality team analyzes the chart and determines what is the process problem regarding the one out-of-control subgroup (subgroup 2), the chart is revised (Figure 6.9). With subgroup 2 removed from the data, the total number of subgroups is 39. Keeping Figure 6.5 handy, the revised calculations become:

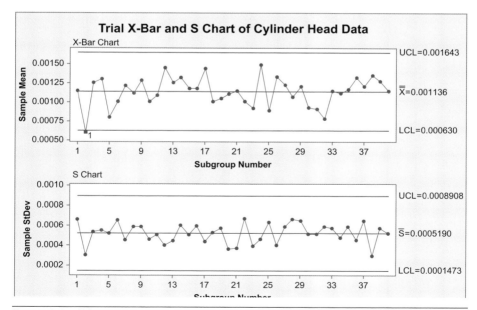

FIGURE 6.8  X-Bar and S chart of cylinder head data

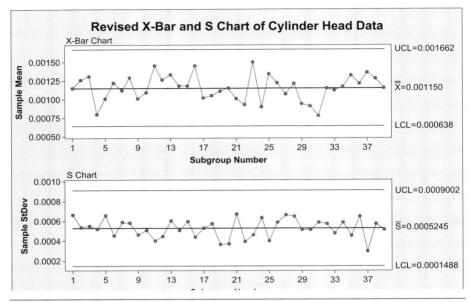

FIGURE 6.9   Revised X-Bar and S chart of cylinder head data

$$\bar{\bar{X}} = \Sigma \bar{x}/n \qquad (2.1)$$
$$= 0.4485/39 = 0.00115$$

$$\mathrm{UCL}_x = \bar{\bar{X}} + A\sigma_o \qquad (6.16)$$
$$= 0.00115 + 0.949(0.0005392) = 0.001662$$

where $\sigma_o$ is the estimated sigma using $s_o$ (0.000524) and $s_o = \bar{s}$ revised (with problem subgroups removed). $\sigma_o$ is computed using equation 6.21 as follows:

$$\sigma_o = s_o/c4 \qquad (6.17)$$
$$= 0.0005245/0.9727 = 0.0005392$$

$$\mathrm{LCL}_x = \bar{\bar{X}} - A\sigma_o \qquad (6.18)$$
$$= 0.00115 - 0.949(0.0005392) = 0.000638$$

$$\bar{s} = \Sigma s/n \qquad (6.19)$$
$$= 0.020456/39 = 0.0005245$$

$$\mathrm{UCL}_s = B_6\sigma_o \qquad (6.20)$$
$$= 1.669(0.0005392) = 0.0009002$$

and

$$LCL_s = B_5 \sigma_o \quad (6.21)$$
$$= 0.276(0.0005392) = 0.0001488$$

The reader should begin to see a pattern emerging among all of the SPC charts in Chapters Five and Six. Although calculations are different with regard to chart type and whether or not the chart is trial or revised, the purpose and procedure remain the same. Until now, analysis was limited to finding points beyond the control limits (an outlier) and searching for problem patterns. These sources of assignable variance were controlled and the process improved. Chapter Seven continues this strategy by digging deeper into chart analysis or what is better described as *process analysis*.

## Chapter Six Discussion

1. Give some examples of continuous variables as opposed to discrete measures or attributes.
2. How can continuous measures be converted to discrete or attribute measures?
3. Discuss the significance of the decimal place and accuracy of measure.
4. What are the three most common variable charts, and how do they differ from attribute charts?
5. What are the steps in creating a variable chart?
6. How is a rational subgroup determined?
7. Why does the centerline of the means have a double bar, but the centerline of the ranges has a single bar?
8. Why is it best practice to use standard SPC chart constants to calculate chart limit values rather than simply computing $\pm 3\sigma$?
9. How do the points on the charts indicate natural or assigned variance?
10. Over time, how will these charts change?
11. Why is close to average in an X-Bar chart desirable, but close to the LCL in an R Chart desirable?

## Chapter Six Problems

1. Use the Excel® and Minitab® data provided for this problem (see Appendix E) and go through the steps to create an SPC study (hint: use every 4 or 5 numbers in the data to represents one subgroup). Tabulate, calculate, and graph the chart to better understand the process. Use both Excel® and Minitab® to gain experience in different computer programs. Note how the chart changes from the trial chart to the revised chart.

2. Find a group of objects around the home or at work that have a continuous measure as a quality characteristic. Go through the steps in creating a SPC study. Using real data from a production line would be best for this exercise. Tabulate, calculate, and graph the chart to better understand the process. Use both Excel® and Minitab® to gain experience in different computer programs. Note how the chart changes from the trial chart to the revised chart.
3. Using the same data from the problems above or collecting new data, repeat the problem. However, compute only an S-Chart in conjunction with the X-Bar. Note any differences. Remember to increase the subgroup size.

# SEVEN

## Control Chart Analysis

Collecting data for control charts displays trends and processes in application. Largely, these charts describe what is happening during the operations level; they provide an opportunity to make decisions regarding control and improvement. However, the charts covered in Chapter Six are basic in nature and do not take into consideration some common variations presented by using these charts. This chapter discusses some of the more realistic aspects of the control chart such as variable subgroup size, type I and II errors, chart analysis with zone rules (and their violations), and process capability including specification, cases of tolerance, and the $C_p$ and $C_{pk}$ statistics.

## Variable Subgroup Sizes

As discussed in Chapters Five and Six, subgroup size is normally contingent upon the measure, the chart, the amount of product available to measure, and the time or expense involved in recording data. For variable charts, the subgroup size is typically set to conventional values of 4 or 5 for X-Bar and R charts and higher for S Charts (10 or more). Attribute charts generally require estimating the proportion that are non-conforming and then setting the subgroup size to assure adequate non-conforming samples to reveal process variance. It is not uncommon to see subgroup sizes in the hundreds or even thousands for attribute charts (with the exception of the c-chart, where there is only one subject per subgroup). In most cases, the subgroup size, once set, remains constant throughout the study. However, in some instances the subgroup must vary for a myriad of reasons, but typically because of 100% inspection or batches sizes that vary in size.

There are three common methods of creating variable subgroup size control charts. They are similar between variable charts and attribute charts. Understand, however, some charts such as the *np*- and *c*-charts do not facilitate a varying subgroup size. The sample size follows an inverse proportion relationship when determining the control limits. As the sample size decreases, the control limits widen. As the sample size increases, the control limits tighten.

What is common between these methods is the task of representing the control limits as per the subgroup size for each of the subgroups in the chart. The regular convention is to select the method and then present the chart in a way that eliminates as much confusion as possible. The method would likely be determined by the manager, who has a comprehensive knowledge of the operative employees and their level of understanding charts.

These methods differ in presentation only. One method simply plots each control limit for every subgroup. To the operative employee, this may present the incorrect notion that limits change daily. A few minutes of training with most employees would overcome confusion in this regard (Figure 7.1).

Another method averages the control limits so that only two limits are found in non-variable sample size control charts. This type of chart is typically better received among the operative employees because its limits are clear. But this method presents its own complications (Figure 7.2). Although the chart has only one control limit — much like a chart with a constant subgroup — out-of-control situations are more difficult to detect.

To rectify this, out-of-control situations are based on four cases. These cases are based on the relationship between control limits and subgroup size. If the subgroup falls within the averaged limits, and the actual subgroup size is smaller than the average subgroup size, the subgroup is within control. This is because the control limits will only become wider as the subgroup size becomes smaller. This is case I and no

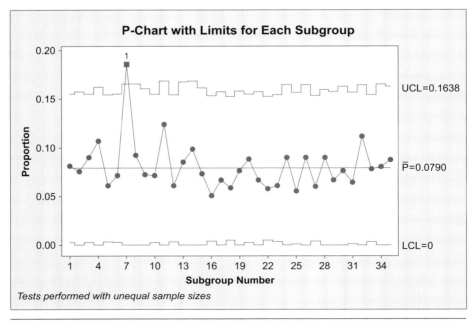

FIGURE 7.1  Variable subgroup size with each control limit

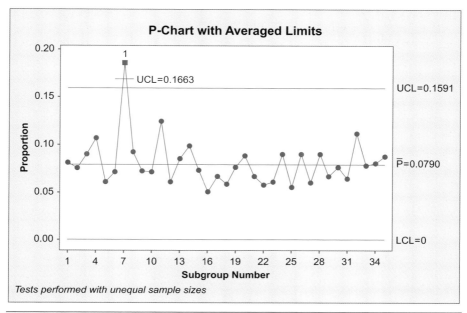

FIGURE 7.2  Variable subgroup size with averaged control limit

action is needed. In Figure 7.2, the averaged UCL of 0.1591 represents an average subgroup size of 102. Approximately half of the subgroups have subgroup sizes larger than the average. Subgroups 4, 7, 8, 9, 11, 13, 14, 15, 24, 26, 28, 30, 32, 34, and 35 may require additional consideration. The remaining subgroups have subgroup sizes less than the average; hence, they fit into case I with no further action.

Case II is where the subgroup falls inside the averaged limit, but the subgroup size is larger than the averaged subgroup size. As this condition tends to narrow the limits closer to the centerline, we need to determine if these subgroups actually in control, even though they may fall within the limits based on average subgroup size. A quick scan of a chart with limits calculated for each subgroup (Figure 7.1) will reveal if the subgroup is within or outside the control limits. In this case, there are no subgroups with higher-than-average subgroup sizes that are close to the control limits.

Next, even if a subgroup is outside of the average limits, it is not necessarily out of limits. In case III, a subgroup outside the average limits is larger than the averaged subgroup size. In this case, the subgroup is indeed out of limits because the true limits are tighter than the average limits. However, if the subgroup size is smaller than the average subgroup and is outside of the averaged limit, the subgroup may be out of control because the true control limits would be wider. This is case IV. Calculating the true control limits will reveal whether or not the subgroup is within limits. This is the case with subgroup seven in Figures 7.1 and 7.2.

The third common method for creating the charts assumes the subgroup sizes are not randomly variable, but follow a few set levels (Figure 7.3). Suppose a plant

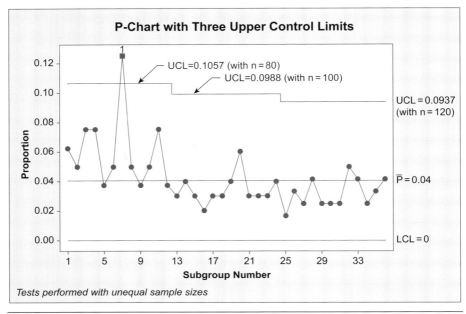

FIGURE 7.3  Variable subgroup size as set levels

routinely manufactures 5000 items on Mondays and Tuesdays, 7500 on Wednesdays and Thursdays, and 10,000 on Fridays and Saturdays (closed on Sundays). There may be three levels of control limits calculated in the control chart, each representing the subgroup size for the respective production level. To quickly ascertain if a subgroup is in control or out of control, the subgroup labeling should indicate which day of the week it is taken.

These are not the only methods for treating variable subgroup size, but they are the ones most commonly found in industry. Again, the primary purpose is to eliminate as much confusion as possible so the information can be presented and universal progress toward improvement continues.

Incidentally, subgroup seven, which has repeatedly been determined to be outside of the upper control limit, should receive SPC treatment. The variability contributing to this outlier will likely be determined and a revised chart then created.

Similarly, this same procedure applies to variable charts such as X-Bar and R charts as shown in Figure 7.4. Note subgroups 12 and 13 in the R chart. Subgroup 13 is out of control. Meanwhile, subgroup 12 is in control, even though the range average for subgroup 12 is at a higher level than subgroup 13. This example is just one of many where these charts with varying subgroup sizes are more difficult to understand and analyze. Unless there is an unavoidable situation that requires a variable subgroup size, every attempt should be made to equalize subgroup size.

Other charts covered previously may not be able to have variable subgroup sizes. The *np*-chart is one example because it shows an actual number of non-conforming

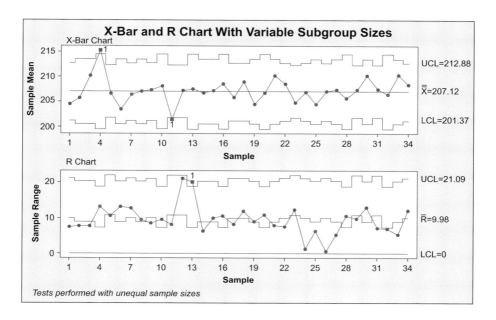

FIGURE 7.4  X-Bar and R chart with variable subgroup sizes

parts. With a variable subgroup size, this number is meaningless. Another example is the c-chart where there are no subgroups and only one item is examined for each point in the chart.

## Process Analysis

Building on Chapters Five and Six, control limits theoretically represent three standard deviations above and below the mean of the subgroups (centerline). By following this empirical rule, most decisions based on the control chart will be sound. However, as with any estimation, there is the possibility of error, even when all assumptions have been met to assure a random, non-biased, normal study. Decisions based on these estimations, which often involve committing financial resources and valuable time, must be tempered with the possibility of an error.

The first error, called a Type I error, blames the non-conformance on an assignable cause when, in fact, the cause was random. Although a point outside of the limits typically means something caused it to be so, it can occur randomly 0.27% of the time. This amount represents the probability of a point being beyond the 3rd standard deviation from the mean. We can conjecture that the farther along the tail of the curve outside of the 3rd standard deviation, the less likely you would make a Type I error (the presence of an outlier). Careful consideration and constant observation of the process must accompany any analysis of a point outside of the limit, and particularly those close to the limits, before assigning a cause.

The second error is a Type II error; it is somewhat the opposite of a Type I error. Normally, and rightfully so, a point inside three standard deviations from the mean is considered conforming and is attributed to random causes. Occasionally, however, a point within the limits may exist from an assignable cause. Sometimes these cases can be detected with short-term trend recognition, but many times they remain unnoticed.

With the understanding of a possibility of error, the next step in chart analysis is to determine if a subgroup or group of subgroups is out of control. Two major scenarios represent an out-of-control condition. First, a single point is outside the control limits. This condition has already been examined several times in this text. Second, a non-conforming pattern exists within the control limits. This non-conformity might be justified based on the normal distribution. For example, if two points are a certain distance away from the mean (measured by standard deviation), the pattern may still remain within the probability of that occurring.

More than two, however, may go beyond the chances of occurring. Because data are to remain random and produce an evenly scattered pattern throughout the chart, any trend or conglomeration of data becomes suspicious. Six zones, three on the positive side of the centerline and three on the negative side, exist to allow detection of abnormal patterns. What are called *zone rules* apply to detect out-of-control conditions within possible abnormal patterns. These zones are typically labeled zone C through A; they correspond to the respective standard deviations from the mean. Sometimes the zones are called one sigma zone, two sigma zone, and three sigma zone (Figure 7.5). Zone C refers to one standard deviation on either side of the centerline of the chart.

Zone rules can be set by the technician. But the following scenario is common. There can be no more than seven points in a row residing within a single zone C or beyond, either above or below the centerline. If this occurs, the technician should suspect an assignable cause. Likewise, there must be no more than four out of five subgroups in a row in zone B or beyond, and two out of three subgroups in a row in zone A. In addition, six points in a row steadily increasing or decreasing constitutes an out-of-control situation. By sight, it is sometimes difficult to spot these rule violations, but most statistics software has built-in functions that detect them and alert the technician.

Be aware that the out-of-control situations discussed here are from a statistical view. In some instances, an out-of-control situation may, from a process perspective, be a good thing. For example, suppose there were six points in a row decreasing toward the mean (or toward the lower limits in some charts), then hovering within zone C above and below the mean. A savvy technician may want to find out why the process suddenly improved and try to repeat those conditions.

Outside of the zone violations, look not only for short-term patterns or trends; also look for long-term patterns or trends in the chart. Any deviation from a random, evenly scattered pattern is a sign that assignable (or unnatural) causes are present. With some investigation, assignable causes are accounted for and may be used by either elimination or duplication to improve the process. Consider Figures 7.6 and 7.7 as examples of how patterns exist. Figure 7.6 shows an upward pattern in the data.

## Control Chart Analysis 125

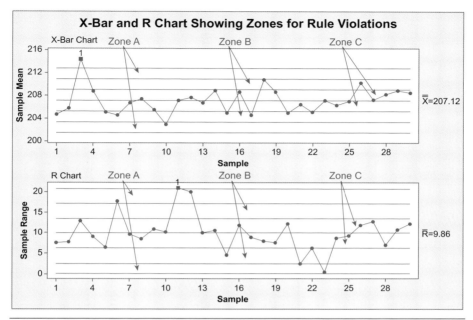

FIGURE 7.5  Zones C through A and respective rule violations

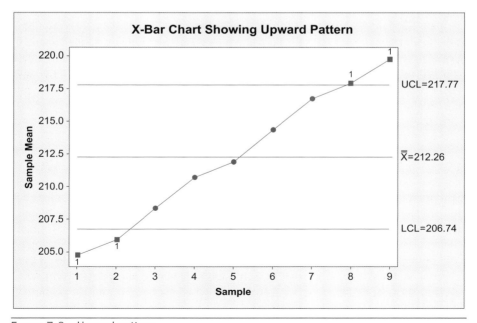

FIGURE 7.6  Upward pattern

**126** Chapter Seven

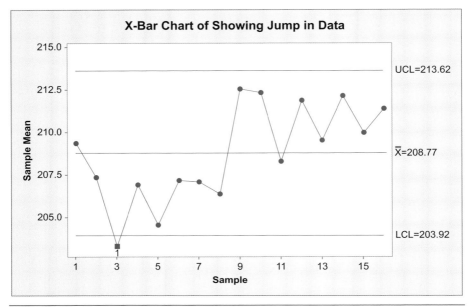

FIGURE 7.7   Jump in data

Figure 7.7 shows a jump in the data; note the elevated position of subgroups 9 through 16 compared to 1 through 8. Because these patterns could be attributed only to assignable cause variance rather than natural variance, what could have caused them? Remember, a thorough documentation during the data collection phase can be invaluable in revealing a cause for any number of patterns and trends.

## Process Control Limits, Tolerance Limits, and Specifications

In Chapters Five and Six, the discussion revolved around creating control charts and understanding that these charts were to describe and control the process. A distinction was drawn between the distribution of individual items manufactured (no subgrouping) and the distribution resulting from sub-grouping those items. An individual distribution is larger than a sub-grouped distribution, but the mean is equal. This difference is critical as different control limits are determined. So far in this chapter, the discussion has centered around the process control chart limits. These are the Upper Control Limit and the Lower Control Limit. They monitor the quality of the process and subsequently the product. A subgroup outside of the control limit does not necessarily mean all items in the subgroup are bad. Individual items must be tested for conformance on a different scale — that of the Upper and Lower Natural Tolerance Limits (UNTL and LNTL). These limits are the 6 sigma (6σ) associated with the individual distribution.

Suppose an aircraft ready for takeoff is centered at one end of the runway. The pilot must follow the centerline while speeding down the runway to rotate and lift off. That centerline is analogous to the centerline in the control chart. As the plane moves down the runway, the pilot is making slight but constant adjustments to the rudder to keep the plane on the centerline. Does it deviate from the centerline? Yes. If there is no deviation, the pilot and the equipment are perfect. Deviation (swerving aircraft) becomes critical if the runway is very narrow.

The edge of the runway, then, can be thought of as the Natural Tolerance Limits (NTLs). Outside of these limits, the plane is off the runway into the grass and runway lights. But the pilot does not wait until they are too close to the edge before correcting the direction. Again, the pilot will make small and constant adjustments as necessary to keep the plane comfortably away from the NTLs. The limits of these constant adjustments can be thought of as the control chart limits. Even if the runway is wide enough, if the pilot has trouble doing this, two things may be at fault. The pilot (operative employee) may not have the skills required to fly that plane (operate that machine), or the plane (process equipment) may not be capable of staying on a straight line (ineffective or antiquated equipment). The solution is to train the operative worker to increase skill, to upgrade the equipment, or both. Each of these scenarios cost money, especially upgrading equipment.

There is another limit to discuss: the specification limit (SL). The specifications determine when a product will not function the way it was designed to function. For example, if the product is a shaft, the diameter may be too large or too tight to fit into the hole it was designed for. Typically there are two specifications: an Upper Specification Limit (USL) and a Lower Specification Limit (LSL). If the machinery is sophisticated enough and the skill of the operative workers is high, the items manufactured may be well away from the specification limits.

This is not necessarily an ideal situation. If the specifications are too far away from the individual limits, too much money is spent purchasing unnecessarily sophisticated equipment. Instead, the ideal is to have the ability to have the NTLs a little lower than the specification limits. Some processes are such that the NTLs are equal to the specification. This is the case of the airplane on the runway. The edge of the runway is the specification limit. If the airplane runs off the runway on the way to takeoff, the flight is likely over! As predicted, there is also the case where the specifications are lower than the NTLs. Although production is possible, there will always be some waste. Figure 7.8 shows each of these three sets of limits as they relate to an individual distribution and a subgrouped distribution. In practice, the SLs are not always in the position as depicted in the figure.

Individual distributions are also useful in creating what are called individuals charts. These charts enable an analysis of product that other charts may not be able to provide. Some of these charts will be discussed later in this chapter.

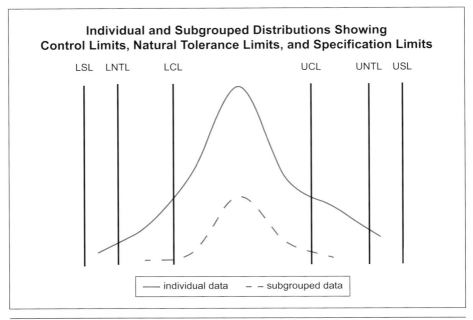

FIGURE 7.8  Individual and subgrouped distributions with control, natural tolerance, and specification limits

## Process Capability

Just as the NTLs are measured in sigma ($\sigma$) for ease of computation, so are the specification limits. Suppose the LSL is 20 millimeters, the USL is 100 mm, the LNTL is 30 mm, and the UNTL is 90 mm. The NTLs would then be $6\sigma$ and the specification limits would be $8\sigma$. (Note the difference in Figure 7.9.) If calculated: 8/6=1.33. Incidentally, this value represents the industrial standard for what is called the *process capability* and is identified by the statistic $C_p$. The $C_p$ equation is as follows:

$$C_p = (USL - LSL)/6\sigma \tag{7.1}$$
$$C_p = 8/6 = 1.33$$

For the data similar to the cylinder head data generated in Chapter Six, this analysis shows how capable the process is in producing the product (Figure 7.9). Three scenarios, referred to as Case I, Case II, and Case III describe the process capability more specifically (Figure 7.10). If the NTLs are lower than the specifications (Case I), the $C_p$ is greater than 1.00. In other words, the process is fully capable of producing the item with room to spare. If the NTLs are equal to the specifications (Case II), the $C_p$ is 1.00 (i.e., $6\sigma/6\sigma = 1.00$). In this case, there is no

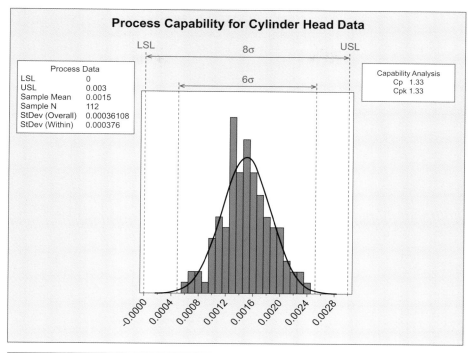

FIGURE 7.9   $C_p$ of NTLs = 6σ and SLs = 8σ

room for error. If the process shifts to one side or the other, there will be waste. If the process continues on center, product will remain within specification. If the NTLs are greater than the specifications (Case III), the $C_p$ is less than 1.00 (i.e., 4σ/6σ = 0.6667). Even if the process is centered, there will be waste. The $C_p$ statistic shows if a process is possible. It shows if the individual distribution will fit within or around the specification limits. It does not account for deviation from the center.

Another useful statistic for process analysis is the $C_{pk}$. The $C_{pk}$ statistic is used to reveal the centering of the process and is sometimes referred to as the *Capability Index*. Not all cases require the process to be centered; in some cases, the process may be purposefully brought off center to maximize production. Usually this happens because the specifications are lower than the NTLs. For example, if the product distribution is centered with a $C_p$ of 0.5, a quarter of the product will be too big, and a quarter of the product will be too small. The product that is too big can be reworked to fit within specifications. The product that is too small, however, has to be scrapped. By shifting the process center toward the top, more reworkable product that is too large is produced, therefore, reducing scrap. In this case, the $C_{pk}$ will not equal the $C_p$. Only if the process is centered will the $C_{pk}$ equal the $C_p$.

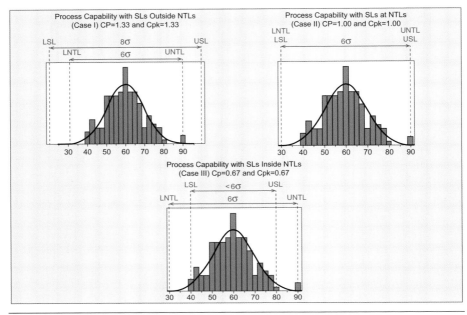

FIGURE 7.10  Three cases of the SLs position relative to the NTLs position

To calculate the $C_{pk}$, think of a process that is off center (Figure 7.11). The area from the mean of the process distribution extending to the specification, both on the positive side and the negative side, can be measured with a z-score. That z-score is compared to half of the process, which is $3\sigma$. The $C_{pk}$ resulting from equation 7.2 is a statistic similar to the $C_p$ that represents how centered the process is. The closer the $C_{pk}$ is to the $C_p$, the closer the process is to being centered. Again, a perfectly centered process will have the same $C_{pk}$ as the $C_p$. Analytically:

$$C_{pk} = z(\min)/3 \qquad (7.2)$$

or

$$C_{pk} = \min(C_{pu}, C_{pl}) \text{ (see explanation below)} \qquad (7.3)$$

where min is the lesser of

$$z(\text{positive}) = (\text{USL} - \bar{x})/\sigma \qquad (7.4)$$

or

$$z(\text{negative}) = (\bar{x} - \text{LSL})/\sigma \qquad (7.5)$$

Control Chart Analysis  **131**

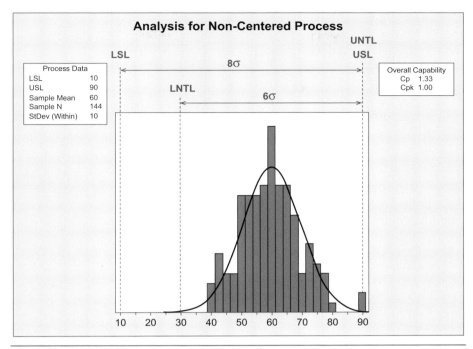

FIGURE 7.11  Shaft example showing positive off-center process

The assumption of normalcy is a must for calculating $C_{pk}$. On occasion, $C_{pl}$ (lower) and $C_{pu}$ (upper) are calculated where $C_{pl}$ is $(\bar{x} - \text{LSL})/3\ \sigma$, and $C_{pu}$ is $(\text{USL} - \bar{x})/3\ \sigma$. Then the $C_{pk}$ becomes the lesser of the two.

Consider a process making shafts to hold a mating part of a machine in a very large power generating application. The process technology is quite new and capable of producing product well within the specification given by the engineers. The shaft diameters produced are 4.003 inches on the lower side and 4.009 inches on the upper side of the distribution. The specifications call for anything in between 4.001 inches and 4.009 inches. The range of the process distribution is 0.006 and is normal. Recognizing only thousandths of an inch, the standard deviation becomes 10 and the mean is 60 (see Appendix E). Using equation 7.1, the $C_p$ is:

$$C_p = (\text{USL} - \text{LSL})/6\sigma$$
$$= (90 - 10)/6 = 8/6 = 1.33$$

As stated, the process is quite capable with a $C_p$ of 1.33. Glancing at the UNTL shows no out of specification product, but any more of a shift toward the positive would produce waste. Calculating the $C_{pk}$ will help to fully grasp this. Using equation 7.2:

$$C_{pk} = z(\min)/3$$

where, knowing the process is on the positive side of the specifications, the minimum z-score will be on the positive side, not the negative side. Using equation 7.4:

$$z(\text{positive}) = (\text{UNTL} - \bar{x})/\sigma$$
$$= (90 - 60)/10 = 3$$

and

$$C_{pk} = 3/3 = 1.00$$

The $C_{pk}$ of 1.00 compared to the $C_p$ of 1.33 confirms the process is not centered. Figure 7.11 shows it is offset to the positive side and flush to the USL of 4.009. Given the sophistication of the machinery, one course of action may be to control the process so the centerline of the chart shifts down to about 4.005. This would result in a $C_{pk}$ equal to the $C_p$ (1.33), and would avoid the expense in rework or scrap if the process shifted to the positive side. The alternative would be do nothing and maintain a tight control on the process to produce the mean of 4.006. In this Case I example, the NTLs are less than the specifications required. Consider the cases depicted in Figure 7.10 along with a non-centered process where $C_p$ and $C_{pk}$ are not equal. What course of action is involved in each of these?

$C_p$ and $C_{pk}$ are overwhelmingly the most common process capability statistics used in industry. However, there are other statistics similar to them, including $C_r$ (Capability Ratio), $P_p$ (Process Performance), and $P_{pk}$ (Process Performance Index). $C_r$ is sometimes calculated as the inverse of $C_p$:

$$C_r = 1/C_p$$

Both are used interchangeably with some confusion. Those involved in the study should clearly define which form of capability statistic, either the $C_p$ or the $C_r$, is used to avoid any such confusion.

Normally, $P_p$ and $P_{pk}$ are interchangeable between $C_p$ and $C_{pk}$, but there is considerable confusion regarding the definition and use of these statistics. $C_p$ and $C_{pk}$ use standard values of R-Bar and $d_2$ or the average of the subgroup standard deviations (Chapter Six) for their calculations. In contrast, $P_p$ and $P_{pk}$ make calculations based on the individual distribution (non-subgrouped) standard deviation. Sometimes this is distinguished by using *within* and *between* terminologies — *within* refers to $C_p$ and $C_{pk}$ whereas *between* refers to $P_p$ and $P_{pk}$. In this sense, $P_p$ and $P_{pk}$ could indicate a potential capability for the process, which in a culture of improvement is analogous to performance.

When the individual distribution standard deviation is used to calculate $C_p$ and $C_{pk}$, $C_p$ is equivalent to $P_p$ and $C_{pk}$ is equivalent to $P_{pk}$. Nevertheless, $C_p$ and $C_{pk}$ commonly refer to process capability whereas $P_p$ and $P_{pk}$ refer to process performance. To reiterate, the most common process statistics continue to be $C_p$ and $C_{pk}$.

This discussion focuses on variable control charts. Attributes require the same analysis, and much of this discussion applies. However, $C_p$, $C_{pk}$, $P_p$, and $P_{pk}$ do not apply to attribute charts. The process capability in attribute charts are built in, meaning the expected value (historical proportion or average) is the process capability and follows a binomial distribution. Continual process improvement over time will show in the improvement of the process capability.

## Individuals Charts

Much of the discussion in this chapter has centered around individual distributions in comparison to or in conjunction with subgrouped distributions. At this point, it is worth pointing out there are some SPC charts that start by utilizing the individual distribution. What follows is a short description of two of the most common individuals charts: the I-MR Chart and the Run Chart.

I-MR Charts (Individuals and Moving Range Charts) appear similar to the X-Bar and R Charts. However, they do not include subgrouping. By recording each individual, the chart can easily map the progress of implemented improvements. The Moving Range part of the chart represents the difference in range (variability) from the previous individual item. The Individual part of the chart represents the average. These charts are quite sensitive to any changes in the process; as such, they can help monitor the system for any type of assignable causes. They are frequently used to determine if a process is stable prior to effecting an improvement. If the process is not stable, the quality team will not be able to determine if the improvement treatment is changing the process or the process was not stable prior to the treatment. I-MR Charts are also used to create a before and after comparison to ascertain the effectiveness of any type of improvement treatment.

Figure 7.12 shows an I-MR Chart for the shaft diameter data example. Note the out-of-control observation number 6 in the MR section. Documentation becomes even more critical with a chart such as this. If no reason is written or observed for this out-of-control part, the process would have to continue until stable. The reason may never be known. By focusing on the process and determining a reason for the out-of-control part, the process will become stable sooner in preparation for the anticipated improvement.

Run Charts simply plot a series of individual values, which may or may not include subgrouped information, over a period of time to quickly see if the process is improving or deteriorating. It is quite common to use this type of chart in a startup process until sufficient data are generated to maintain charts with centerlines and limits. Run Charts allow an instant visualization of patterns to help determine where problems exist with the process; they can facilitate several analytic tests for trends, patterns, and other special cause variation. Run charts can also help determine the expected value or ppm (parts per million) of nonconforming parts and give a basis for historical data. Figure 7.13a shows an example of a Run Chart with individual observations. It gives calculated probabilities for trends and patterns along with some data about runs. Figure 7.13b shows an example of a Run Chart using both subgroups and individuals for one particular machine over a period of time. In this chart, the reference line in the middle represents the median.

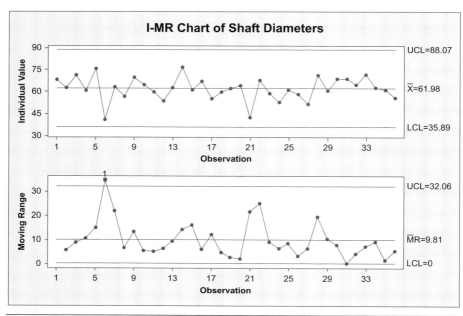

FIGURE 7.12   I-MR chart of shaft diameter data

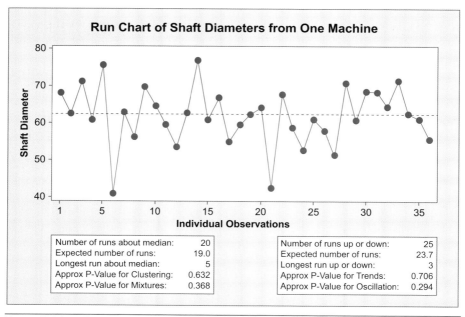

FIGURE 7.13a   Individual run chart of shaft diameter data

# Control Chart Analysis 135

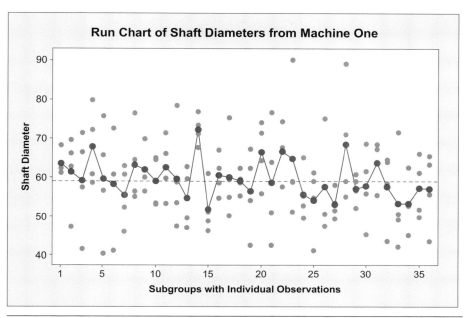

FIGURE 7.13b   Subgrouped run chart of shaft diameter data

## Chapter Seven Discussion

1. Why would an SPC study have a variable number of data in the subgroup?
2. What are the graphic methods of representing variable subgroup size? How would an out-of-control condition be determined?
3. Describe the difference between a Type 1 error and a Type II error.
4. What are the rule violations and how do they help determine an out-of-control condition? Can these rule violations be customized?
5. Describe some of the patterns seen in SPC charts and give some possible explanations. With each explanation, speculate on ways to correct these patterns if required.
6. What is the difference between control limits, tolerance limits, and specification limits?
7. Give examples of a 0.67 $C_p$, a 1.00 $C_p$, and a 1.33 $C_p$ and describe the three cases of process capability.
8. What is the distinction between a $C_p$ and $C_{pk}$? When will the $C_p$ and $C_{pk}$ be the same?
9. When will an I-MR Chart and Run Chart be used?

## Chapter Seven Problems

1. Use the data set provided for this problem with Minitab®, Excel®, or both (see Appendix E). Create a variable control chart with a variable subgroup size. Choose and utilize an appropriate graphing method to accommodate the variable subgroup size.

2. Look for outliers and rule violations within the control chart generated in Problem 1. Signify what may cause the out-of-control conditions and any patterns that are present.

3. Calculate specification limits for Problem 1, assuming a $C_p$ of 0.67, 1.00, and 1.33.

4. Shift the distribution in Problem 1 to locate a $C_{pk}$ of 1.00. Experiment with the minimum and maximum z to see how this affects the $C_{pk}$.

# EIGHT
## Acceptance Sampling and Inspection

SPC is an excellent system for monitoring and improving the process quality and, hence, the product itself. The final product, however, may not be made entirely of components produced on site where process monitoring is possible. It is essential that operations assure the high quality of all constituents of the product. Making something in industry requires energy, equipment, labor, and materials. Each is as important as the other because production does not occur with the absence of any one. Equipment and labor are fixed and permanent in regard to the process. Energy and especially materials are not. Although energy quality is an issue and methods discussed in this chapter apply to energy, it is not a direct component of the product, but of the process. This chapter emphasizes the quality of incoming materials and subcontracted subcomponents that combine through the process to form the product. Logically, this system also applies as a monitor for production, especially when product is shipped as material or subcomponents to an industrial consumer for further process.

Normally the expense in measuring 100% of product received from outside sources is prohibitive on a continual basis. To limit this expense, acceptance sampling ascertains and maintains the quality of incoming items based on a sample of the entire lot. Although plans for sampling accommodate variable measures, most generate attribute data. Even studies requiring variable measures commonly view the subject as either conforming or non-conforming. As a result, the statistics used in these studies, and the subsequent standardized tables, are based on binomial, Poisson, and sometimes hypergeometric probability distributions.

## Sampling

This first task in assuring valid results in any acceptance plan is securing a non-bias, random sample with a predetermined number of items to assure reliability. As discussed in Chapter Two, the ideal sampling will show only natural variance or unassignable causes. Any manipulation of subjects or outside influence during a sampling will result in unnatural variance, also referred to as assignable causes. I recommend a brief review of Chapter Two regarding sampling before continuing.

## Randomness

Randomness is affected by selecting too many samples from a similar location, one part of a lot, or any other method that puts too many items within a similar group. Think of polling a group of people to predict who the next president will be and asserting that this poll's results represent America. If the sample is selected entirely in Alabama, the next president will most likely be Republican. If the sample is taken entirely in Hawaii, the next president will instead most likely be Democrat. Restricting the sample to either state represents results expected with that state only, not the entire country. Concentrating on rural areas as opposed to cities results in similar problems, as would many other conditions.

To achieve randomness, this study requires careful examination of demographics and a proportionate representation of those demographics within the poll. With a certain number of observations found by random selection within each demographic, the resulting study gets closer to representing America. Selecting random items from a lot is much simpler than the example above. But, in both cases, randomness is planned into the study by determining where and how observations occur.

One common method for planning randomness is using serial numbers. A random number chart can match all or a part of a serial number used to identify an item for inspection. If serial numbers are not available, some sort of mapping the lot with areas will work. Find the specific area with a random number and then select an item from that area.

For example, picture a barrel of molasses as a single lot (Figure 8.1). Some of the quality measures to determine acceptance may include viscosity, opacity, sugar content, and color. To map the barrel, we might stratify the barrel into sequential locations. Vertical, horizontal, or depth segments in inches or millimeters can find a location anywhere within the barrel, such as A3 or D4. Random numbers (with ranges congruent with the dimensions of the barrel) give corresponding $x$, $y$, and $z$ coordinates. With a small scoop on a long handle, the technician can locate the $x$, $y$, and $z$ coordinates of the barrel, and then sample that random area. Another method, again using random numbers, locates the $x$ with the vertical measure within the barrel, using a radial and distance from the center of the barrel for the other coordinate. Either way, this approach finds a random area within the barrel for sampling.

Sampling any product in industry presents a challenge in finding a way to gather data in a random, consistent manner. As with the barrel example, there may be several methods from which to choose.

## Bias

The line between randomness and bias is blurry. All bias produces a lack of randomness. But not all lack of randomness is bias. Bias can be either unintentional or intentional. When a bacon packaging plant places the best pieces in the cellophane window of the package and the pieces under it are not of the same quality, it amounts to intentional bias through misconception. Making a judgment about the entire lot of bacon by looking at the first piece would lead to accepting the lot when it should have been

## Acceptance Sampling and Inspection 139

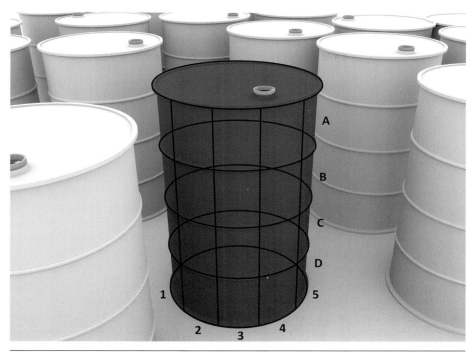

FIGURE 8.1 Stratification of a barrel of molasses

rejected. Intentional bias can take on more serious forms that lead to harm and expense. If incoming lots are doctored by placing good product where inspection will most probably occur, process machinery can be damaged; product quality then deteriorates.

Unintentional bias occurs when someone subconsciously steers the finding of a study in a certain direction– most likely toward their initial hypothesis. Suppose several lots have recently been rejected from a producer and these incidents have caused some irritation and trouble in from the consumer's perspective. New incoming lots may also be rejected because of prejudice (unintentional bias). On the opposite side, item inspection may be overly positive if there is a close positive relationship with a producer who has a reputation of reliably high quality. In other words, individual inspection is pulled in one direction or another based on past studies — again, this is an example of unintentional bias.

The antidote for bias (including some intentional bias) is to follow a plan of randomness while taking samples, use a standard consistent approach to measurement where everyone follows the same procedure (and hence the repeatable reliability is high), and maintain an objective inspection staff that follows a system of checks and balances for the entire team. Any personnel involved in acceptance sampling should undergo training in all of these areas; they should also develop an awareness of the potential harm and expense of any kind of bias.

## A Sufficient Sampling

In addition to ensuring randomness and filtering out bias, the sample size and rejection (or acceptance) criteria provide a rational procedure to make a decision on incoming lots. With most acceptance sampling, determining the sample size and the various parameters that affect rejection or acceptance is quite straightforward. ISO 2859 is the internationally accepted standard for creating acceptance sampling plans. In the United States, these standards are administered by ANSI/ASQ z1.4 for attribute data and ANSI/ASQ z1.9 for variable data. From the standard charts, inspectors can plan the sample size based on lot size; they can plan the acceptance and rejection numbers based on the Acceptance Quality Level (AQL). Other plans may seek to fulfill certain stipulations as per consumer advocacy or producer advocacy. Description and analysis of these plans are accomplished using an Operating Characteristic (OC) Curve that is negotiated between each party. These concepts and more are discussed below.

## Types of Plans

First, review the common notation for acceptance sampling: N = lot size, n = sample size, and c = acceptance level. More complicated plans go further with r = rejection level. For each subsequent sample, if needed, these criteria change. Another important consideration, the known expected value (sometimes called the mean, or $\mu$) for nonconformity (see Chapter Three), is the historical probability that a randomly selected individual item will be nonconforming.

For the most part, determining sample size in industry has become standardized. Both attribute and variable standards are available through the aforementioned organization. However, due to the complexity of variable acceptance plans, this text will concentrate on the more common attribute acceptance plans disseminated through the ANSI/ASQ z1.4 standard.

### Single Sampling Plans

A single sampling plan is the simplest to explain and has only three parameters: N, n, and c. Lot size (N) is based on the product item; it may be small, large, simple, complicated, or heavy. Using the percent that are nonconforming, a sample (n) provides a measurable number (c) based on the probability to either accept or reject the lot. When the lot comes into the inspection center, the inspection team draws the sample, makes measures of each item (usually attribute), and counts the number nonconforming. If the count of nonconforming is c or less than c, the lot is accepted and used in the plant. If the count of nonconforming is greater than c, the lot is rejected for further consideration by either the producer or the consumer. Many times in practice, the inspection team leaders will simply give instructions to sample a certain percentage of the lot. This works well in most cases. If fewer parts are needed for a valid study, less expense could be involved. On the other hand, if the sample size is too low, the results may be erroneous — meaning a bad lot was accepted or a good lot was rejected. A more stable approach to selecting sample size (n), and the subsequent acceptance value (c), is using standardized tables to find sample size and acceptance values.

These standardized tables were developed by Harold Dodge and Harry Romig in the 1920s while they were employees of Bell Laboratories. Contemporaries and colleagues of Walter Shewhart, Dodge and Romig compiled standards for acceptance sampling and inspection still used today. During World War II, Dodge and Romig adapted their work for the military under MIL-STD-105D. This standardized approach in production contributed to the war effort in making productivity more efficient through higher quality and increased production. At the end of World War II, these standards followed into industry. The MIL-STD-105D was later revised into MIL-STD-05E and then transferred into ANSI/ASQz1.4, and ultimately to the international standard ISO 2859.

Basically, the tables include a sample size code dependent on inspection level and any special circumstances, an Average Quality Level (AQL) that describes the level of nonconforming parts in percentage form (or per 100 units), and a means of determining the acceptance and rejection values (c and r, referred to respectively as Ac and Re in the tables). These standards are still commonly referred to as MIL-STD-105D in industry.

Consider an example. A company fabricates cables used in small aircraft to facilitate the control of flight surfaces. These surfaces include the ailerons, the rudder, and the elevator. One end of the cables is attached to these surfaces and the other end is attached to the yoke (the steering wheel) and the rudder control peddles. As the pilot pulls the yoke forward, the elevator (those little wings on the tail of the plane) responds with an upward motion. As the yoke returns and moves forward, the elevator responds with a downward motion. This motion causes the plane to fly up or down. If the cable fails, there will be a problem controlling that surface and the aircraft will likely meet with an accident.

This company buys several 5000-foot spools of cable, cuts the cable to specific lengths, and attaches the necessary fixtures to the cable for use in the aircraft. They then send the ready-to-install cables to various contracted aircraft manufacturers. When the company receives a spool of cable, the acceptance or rejection process begins. This particular cable has a 1/8″ diameter and a minimum breaking strength of 1700 lbs. of tensile force. One of the tests involves making sure the cables can hold the specified stress. Because the company wants to know only if it will hold a minimum of 1700 lbs., they are not interested in the maximum or any variation between the minimum and maximum. If the cable can hold 1700 lbs., it passes the test. The study now becomes an attribute study with the outcome being either pass (it did hold 1700 lbs.) or fail (it did not hold 1700 lbs.). Of course, failure could mean the cable broke outright or presented some other sign of failure such as broken strands, excessive stretching past the modulus of elasticity, warping, or cracks.

The testing process requires cutting several 12″ portions of the cable out of the spool, attaching each 12″ cable into a tensile tester, and exerting 1700 lbs. of tensile pressure to see if the cable fails. Any cable put through this destructive type test is discarded, which is the reason why 12″ portions were taken rather than testing uncut

TABLE 8.1  Sample Size Code Letter, *Reprinted with permission, ©2013 ASQ, http://asq.org. No further distribution allowed without permission.*

| Lot or Batch Size | General Inspection Levels | | | Special Inspection Levels | | | |
|---|---|---|---|---|---|---|---|
| | I | II | III | S-1 | S-2 | S-3 | S-4 |
| 2 to 8 | A | A | B | A | A | A | A |
| 9 to 15 | A | B | C | A | A | A | A |
| 16 to 25 | B | C | D | A | A | B | B |
| 26 to 50 | C | D | E | A | B | B | C |
| 51 to 90 | C | E | F | B | B | C | C |
| 91 to 150 | D | F | G | B | B | D | C |
| 151 to 280 | E | G | H | B | C | D | E |
| 281 to 500 | F | H | J | B | C | D | E |
| 501 to 1200 | G | J | K | C | C | E | F |
| 1201 to 3200 | H | K | L | C | D | G | E |
| 3201 to 10000 | J | L | M | C | D | G | F |
| 10001 to 35000 | K | M | N | C | D | F | H |
| 35001 to 150000 | L | N | P | D | E | G | J |
| 150001 to 500000 | M | P | Q | D | E | G | J |
| 500001 and over | N | Q | R | D | E | H | K |

portions from the spool. Looking at Table 8.1, testing technicians find the sample size code for this study. The lot size of N = 5000 is in the first column within the values 3201 to 10000. Moving over to General Inspection Levels, under II, they select the sample code L. General Inspection Level II is a normal inspection level. Levels I and III are used to increase or decrease sample sizes based on risk and expense. The Special Inspection Levels in the table allow smaller samples when testing is extremely expensive or dangerous to the inspection team.

Suppose the company, by contract with the cable producer, sets an AQL of 0.10. The producer does what is necessary with the cable-making process to ensure the cable they send to the customer is no more than 0.10 percent defective. If this value is high for this type of process, the producer has some leeway and can possibly save money in relaxing quality. If the value is low, the producer must improve the process, often with new more sophisticated equipment, to ensure the appropriate quality. Under the first two columns of Table 8.2, the inspection team determines they require a sample of n = 200. They commence cutting 200 12″ portions of cable from the 5000-foot spool.

Cutting all 200 pieces from the first 200 feet of the spool will not be a random sample. The team determines a random approach by generating a random number table of 200 groups of 10 numbers each, between 1 and 5000. In the event the top numbers of any two groups are the same, the next number in the group gives an alternative (Table 8.3). Each number represents a particular foot along the 5000-foot spool where the 12″ piece of the sample is removed and labeled. The sample is then tested as per the method indicated above.

Looking again at Table 8.2, move along the row designated by the sample size code letter L, and stop at the column designated by the AQL of 0.10. At that point on the table, there will be an arrow giving direction to the required values, indicating the

Acceptance Sampling and Inspection  **143**

TABLE 8.2  Single Sampling Plans for Normal Inspection. Reprinted with permission, ©2013 ASQ, http://asq.org. No further distribution allowed without permission.

| Sample Size Code Letter | Sample Size | Acceptable Quality Levels (normal inspection) | | | | | | | | | | | | | | | | | | | | | | | | | | | | | | | |
|---|---|---|---|---|---|---|---|---|---|---|---|---|---|---|---|---|---|---|---|---|---|---|---|---|---|---|---|---|---|---|---|---|
| | | 0.010 | | 0.015 | | 0.025 | | 0.040 | | 0.065 | | 0.10 | | 0.15 | | 0.25 | | 0.40 | | 0.65 | | 1.0 | | 1.5 | | 2.5 | | 4.0 | | 6.5 | | 10 |
| | | Ac | Re | Ac | Re | Ac | Re | Ac | Re | Ac | Re | Ac | Re | Ac | Re | Ac | Re | Ac | Re | Ac | Re | Ac | Re | Ac | Re | Ac | Re | Ac | Re | Ac | Re | Ac | Re |

(full AQL table with sample sizes A=2, B=3, C=5, D=8, E=13, F=20, G=32, H=50, J=80, K=125, L=200, M=315, N=500, P=800, Q=1250, R=2000)

Note: If sample size equals, or exceeds, lot or batch size, do 100% inspection.

TABLE 8.3  200 Groups of Random Numbers between 1 and 5000 (partial)

| Between 1 and 5000 (partial) | | | | | | |
|---|---|---|---|---|---|---|
| 3691 | 490 | 3441 | 4000 | 3786 | 4012 | 2637 |
| 4698 | 2400 | 3244 | 1963 | 1199 | 2875 | 3636 |
| 4568 | 1213 | 1623 | 3762 | 4434 | 186 | 4978 |
| 2262 | 2411 | 881 | 4302 | 299 | 3872 | 2399 |
| 1877 | 892 | 3052 | 3334 | 18 | 3631 | 2990 |
| 854 | 187 | 444 | 4667 | 4062 | 235 | 3776 |
| 683 | 2429 | 1066 | 312 | 3845 | 4364 | 298 |
| 4105 | 4222 | 4773 | 3908 | 1584 | 2567 | 1993 |
| 3972 | 2866 | 1595 | 3463 | 2144 | 2398 | 110 |
| 289 | 4670 | 2602 | 4952 | 4302 | 1333 | 4697 |
| 4702 | 1999 | 288 | 4451 | 4976 | 447 | 1685 |

Ac value is 0 and the Re value is 1. If all 200 pieces pass the test, the lot is accepted and production of the aircraft cable assemblies continues. If any of the 200 pieces fail, the lot is rejected and a subsequent agreement between the producer and customer is needed to resolve the issue.

This study could have easily changed with several factors influencing the outcome. First, the General Inspection Level may have been more or less restrictive because of the nature and use of the end line item. However, in this case, the testing criteria (1700 lbs.) was high enough away from the normal operating stress incurred through flight that it was unnecessary to increase the inspection level. Furthermore, based on past studies with product from this producer, the table used could be change from Normal to either Tightened or Reduced. Normal inspection usually occurs during ongoing production. After a specified and agreed upon period with no rejected lots, sampling would change to a reduced inspection to save money and time. In Table 8.4, the sample size with code L reduces from 200 to 80. In the event of a rejected lot, the plan shifts to Tightened inspection. In Table 8.5, the sample size remains at 200 but the AQL = 0.10 column shifts down one row. Although this shift makes little difference in this particular example, we can see how the study is tightened with other AQL values in lower rejection values (Re).

## Double and Multiple Sampling Plans

In the preceding example, only one sample was drawn and inspected from the lot. The acceptance or rejection of that lot rested on whether or not that one sample produced the number of nonconforming units as specified by the appropriate inspection plan table. The ability to accept a lot based on one sample indicates a willing to accept the risk that the sample was valid and representative of the lot. To reduce that risk, studies may incorporate double, or even multiple samples from the same lot. In addition, double and multiple samples can save time and money given a good lot. As the total

TABLE 8.4 Single Sampling Plans for Reduced Inspection. *Reprinted with permission, ©**2013 ASQ**, http://asq.org. No further distribution allowed without permission.*

| Sample Size Code Letter | Sample Size | 0.010 Ac Re | 0.015 Ac Re | 0.025 Ac Re | 0.040 Ac Re | 0.065 Ac Re | 0.10 Ac Re | 0.15 Ac Re | 0.25 Ac Re | 0.40 Ac Re | 0.65 Ac Re | 1.0 Ac Re | 1.5 Ac Re | 2.5 Ac Re | 4.0 Ac Re | 6.5 Ac Re | 10 Ac Re | 15 Ac Re | 25 Ac Re | 40 Ac Re | 65 Ac Re | 100 Ac Re | 150 Ac Re | 250 Ac Re | 400 Ac Re | 650 Ac Re | 1000 Ac Re |
|---|---|---|---|---|---|---|---|---|---|---|---|---|---|---|---|---|---|---|---|---|---|---|---|---|---|---|---|
| | | *Acceptable Quality Levels (normal inspection)* | | | | | | | | | | | | | | | | | | | | | | | | | |
| A | 2 | ↓ | ↓ | ↓ | ↓ | ↓ | ↓ | ↓ | ↓ | ↓ | ↓ | ↓ | ↓ | ↓ | ↓ | 0 1 | ↓ | ↓ | 1 2 | 2 3 | 3 4 | 5 6 | 7 8 | 10 11 | 14 15 | 21 22 | 30 31 |
| B | 2 | ↓ | ↓ | ↓ | ↓ | ↓ | ↓ | ↓ | ↓ | ↓ | ↓ | ↓ | ↓ | ↓ | 0 1 | ↑ | ↓ | 0 2 | 1 3 | 2 4 | 3 5 | 5 6 | 7 8 | 10 11 | 14 15 | 21 22 | 30 31 |
| C | 2 | ↓ | ↓ | ↓ | ↓ | ↓ | ↓ | ↓ | ↓ | ↓ | ↓ | ↓ | ↓ | 0 1 | ↑ | ↓ | 0 2 | 1 3 | 1 4 | 2 5 | 3 6 | 5 8 | 7 10 | 10 13 | 14 17 | 21 24 | ↑ |
| D | 3 | ↓ | ↓ | ↓ | ↓ | ↓ | ↓ | ↓ | ↓ | ↓ | ↓ | ↓ | 0 1 | ↑ | ↓ | 0 2 | 1 3 | 1 4 | 2 5 | 3 6 | 5 8 | 7 10 | 10 13 | 30 31 | 44 45 | ↑ | ↑ |
| E | 5 | ↓ | ↓ | ↓ | ↓ | ↓ | ↓ | ↓ | ↓ | ↓ | ↓ | 0 1 | ↑ | ↓ | 0 2 | 1 3 | 1 4 | 2 5 | 3 6 | 5 8 | 7 10 | 10 13 | 30 31 | 44 45 | ↑ | ↑ | ↑ |
| F | 8 | ↓ | ↓ | ↓ | ↓ | ↓ | ↓ | ↓ | ↓ | ↓ | 0 1 | ↑ | ↓ | 0 2 | 1 3 | 1 4 | 2 5 | 3 6 | 5 8 | 7 10 | 10 13 | ↑ | ↑ | ↑ | ↑ | ↑ | ↑ |
| G | 13 | ↓ | ↓ | ↓ | ↓ | ↓ | ↓ | ↓ | ↓ | 0 1 | ↑ | ↓ | 0 2 | 1 3 | 1 4 | 2 5 | 3 6 | 5 8 | 7 10 | 10 13 | ↑ | ↑ | ↑ | ↑ | ↑ | ↑ | ↑ |
| H | 20 | ↓ | ↓ | ↓ | ↓ | ↓ | ↓ | ↓ | 0 1 | ↑ | ↓ | 0 2 | 1 3 | 1 4 | 2 5 | 3 6 | 5 8 | 7 10 | 10 13 | ↑ | ↑ | ↑ | ↑ | ↑ | ↑ | ↑ | ↑ |
| J | 32 | ↓ | ↓ | ↓ | ↓ | ↓ | ↓ | 0 1 | ↑ | ↓ | 0 2 | 1 3 | 1 4 | 2 5 | 3 6 | 5 8 | 7 10 | 10 13 | ↑ | ↑ | ↑ | ↑ | ↑ | ↑ | ↑ | ↑ | ↑ |
| K | 50 | ↓ | ↓ | ↓ | ↓ | ↓ | 0 1 | ↑ | ↓ | 0 2 | 1 3 | 1 4 | 2 5 | 3 6 | 5 8 | 7 10 | 10 13 | ↑ | ↑ | ↑ | ↑ | ↑ | ↑ | ↑ | ↑ | ↑ | ↑ |
| L | 80 | ↓ | ↓ | ↓ | ↓ | 0 1 | ↑ | ↓ | 0 2 | 1 3 | 1 4 | 2 5 | 3 6 | 5 8 | 7 10 | 10 13 | ↑ | ↑ | ↑ | ↑ | ↑ | ↑ | ↑ | ↑ | ↑ | ↑ | ↑ |
| M | 125 | ↓ | ↓ | ↓ | 0 1 | ↑ | ↓ | 0 2 | 1 3 | 1 4 | 2 5 | 3 6 | 5 8 | 7 10 | 10 13 | ↑ | ↑ | ↑ | ↑ | ↑ | ↑ | ↑ | ↑ | ↑ | ↑ | ↑ | ↑ |
| N | 200 | ↓ | ↓ | 0 1 | ↑ | ↓ | 0 2 | 1 3 | 1 4 | 2 5 | 3 6 | 5 8 | 7 10 | 10 13 | ↑ | ↑ | ↑ | ↑ | ↑ | ↑ | ↑ | ↑ | ↑ | ↑ | ↑ | ↑ | ↑ |
| P | 315 | ↓ | 0 1 | ↑ | ↓ | 0 2 | 1 3 | 1 4 | 2 5 | 3 6 | 5 8 | 7 10 | 10 13 | ↑ | ↑ | ↑ | ↑ | ↑ | ↑ | ↑ | ↑ | ↑ | ↑ | ↑ | ↑ | ↑ | ↑ |
| Q | 500 | 0 1 | ↑ | ↓ | 0 2 | 1 3 | 1 4 | 2 5 | 3 6 | 5 8 | 7 10 | 10 13 | ↑ | ↑ | ↑ | ↑ | ↑ | ↑ | ↑ | ↑ | ↑ | ↑ | ↑ | ↑ | ↑ | ↑ | ↑ |
| R | 800 | ↑ | ↑ | 0 2 | 1 3 | 1 4 | 2 5 | 3 6 | 5 8 | 7 10 | 10 13 | ↑ | ↑ | ↑ | ↑ | ↑ | ↑ | ↑ | ↑ | ↑ | ↑ | ↑ | ↑ | ↑ | ↑ | ↑ | ↑ |

Note: If sample size equals, or exceeds, lot or batch size, do 100% inspection.

Note: If the acceptance number has been exceeded, but the rejection number has not been reached, accept the lot, but reinstate normal inspection.

**TABLE 8.5** Single Sampling Plans for Tightened Inspection. Reprinted with permission, ©2013 ASQ, http://asq.org. No further distribution allowed without permission.

| Sample Size Code Letter | Sample Size | \multicolumn{32}{c}{Acceptable Quality Levels (normal inspection)} |||||||||||||||||||||||||||||||
|---|---|---|---|---|---|---|---|---|---|---|---|---|---|---|---|---|---|---|---|---|---|---|---|---|---|---|---|---|---|---|---|---|---|
| | | 0.010 | | 0.015 | | 0.025 | | 0.040 | | 0.065 | | 0.10 | | 0.15 | | 0.25 | | 0.40 | | 0.65 | | 1.0 | | 1.5 | | 2.5 | | 4.0 | | 6.5 | | 10 | | 15 | | 25 | | 40 | | 65 | | 100 | | 150 | | 250 | | 400 | | 650 | | 1000 | |
| | | Ac | Re | Ac | Re | Ac | Re | Ac | Re | Ac | Re | Ac | Re | Ac | Re | Ac | Re | Ac | Re | Ac | Re | Ac | Re | Ac | Re | Ac | Re | Ac | Re | Ac | Re | Ac | Re | Ac | Re | Ac | Re | Ac | Re | Ac | Re | Ac | Re | Ac | Re | Ac | Re | Ac | Re | Ac | Re | Ac | Re |
| A | 2 | | | | | | | | | | | | | | | | | | | | | | | | | | | | | | | | | | | | | | 1 | 2 | 2 | 3 | 3 | 4 | 5 | 6 | 8 | 9 | 12 | 13 | 18 | 19 | 27 | 28 |
| B | 3 | | | | | | | | | | | | | | | | | | | | | | | | | | | | | | | | | | | 1 | 2 | 2 | 3 | 3 | 4 | 5 | 6 | 8 | 9 | 12 | 13 | 18 | 19 | 27 | 28 | 41 | 42 |
| C | 5 | | | | | | | | | | | | | | | | | | | | | | | | | | | | | | | | | 1 | 2 | 2 | 3 | 3 | 4 | 5 | 6 | 8 | 9 | 12 | 13 | 18 | 19 | 27 | 28 | 41 | 42 | | |
| D | 8 | | | | | | | | | | | | | | | | | | | | | | | | | | | | | | | 1 | 2 | 2 | 3 | 3 | 4 | 5 | 6 | 8 | 9 | 12 | 13 | 18 | 19 | 27 | 28 | 41 | 42 | | | | |
| E | 13 | | | | | | | | | | | | | | | | | | | | | | | | | | | | | 0 | 1 | | | 1 | 2 | 2 | 3 | 3 | 4 | 5 | 6 | 8 | 9 | 12 | 13 | 18 | 19 | 27 | 28 | 41 | 42 | | | | | | |
| F | 20 | | | | | | | | | | | | | | | | | | | | | | | | | | | 0 | 1 | | | 1 | 2 | 2 | 3 | 3 | 4 | 5 | 6 | 8 | 9 | 12 | 13 | 18 | 19 | | | | | | | | | | | | |
| G | 32 | | | | | | | | | | | | | | | | | | | | | | | 0 | 1 | | | 1 | 2 | 2 | 3 | 3 | 4 | 5 | 6 | 8 | 9 | 12 | 13 | 18 | 19 | | | | | | | | | | | | | | | | |
| H | 50 | | | | | | | | | | | | | | | | | | | | | 0 | 1 | | | 1 | 2 | 2 | 3 | 3 | 4 | 5 | 6 | 8 | 9 | 12 | 13 | 18 | 19 | | | | | | | | | | | | | | | | | | |
| J | 80 | | | | | | | | | | | | | | | | | | | 0 | 1 | | | 1 | 2 | 2 | 3 | 3 | 4 | 5 | 6 | 8 | 9 | 12 | 13 | 18 | 19 | | | | | | | | | | | | | | | | | | | | |
| K | 125 | | | | | | | | | | | | | | | | | 0 | 1 | | | 1 | 2 | 2 | 3 | 3 | 4 | 5 | 6 | 8 | 9 | 12 | 13 | 18 | 19 | | | | | | | | | | | | | | | | | | | | | | |
| L | 200 | | | | | | | | | | | | | | | 0 | 1 | | | 1 | 2 | 2 | 3 | 3 | 4 | 5 | 6 | 8 | 9 | 12 | 13 | 18 | 19 | | | | | | | | | | | | | | | | | | | | | | | | |
| M | 315 | | | | | | | | | | | | | 0 | 1 | | | 1 | 2 | 2 | 3 | 3 | 4 | 5 | 6 | 8 | 9 | 12 | 13 | 18 | 19 | | | | | | | | | | | | | | | | | | | | | | | | | | |
| N | 500 | | | | | | | | | | | 0 | 1 | | | 1 | 2 | 2 | 3 | 3 | 4 | 5 | 6 | 8 | 9 | 12 | 13 | 18 | 19 | | | | | | | | | | | | | | | | | | | | | | | | | | | | |
| P | 800 | | | | | | | | | 0 | 1 | | | 1 | 2 | 2 | 3 | 3 | 4 | 5 | 6 | 8 | 9 | 12 | 13 | 18 | 19 | | | | | | | | | | | | | | | | | | | | | | | | | | | | | | |
| Q | 1250 | | | | | | | 0 | 1 | | | 1 | 2 | 2 | 3 | 3 | 4 | 5 | 6 | 8 | 9 | 12 | 13 | 18 | 19 | | | | | | | | | | | | | | | | | | | | | | | | | | | | | | | | |
| R | 2000 | | | | | 0 | 1 | | | 1 | 2 | | | | | | | | | | | | | | | | | | | | | | | | | | | | | | | | | | | | | | | | | | | | | | |
| S | 3150 | | | | | 1 | 2 | | | | | | | | | | | | | | | | | | | | | | | | | | | | | | | | | | | | | | | | | | | | | | | | | | |

Note: If sample size equals, or exceeds, lot or batch size, do 100% inspection.

sample size (from the sample size code letter) is split between double or multiple samples, there is the opportunity to accept the lot based on the first sample in the event the first sample produced at or below the Ac value of nonconformities.

In a double sampling plan, the sample size is determined the same way a single sample plan sample size is determined — by locating a code letter corresponding to the lot size. Inspection level is also determined the same way. Level II would be considered a starting point for normal inspection. Then, using the appropriate inspection plan table for double samples, draw the first sample based of the AQL and the sample size code letter. For example, given a sample size code P and an AQL of 0.10, the first sample requires 500 items. If the sample produces no nonconformities, the lot is accepted (Ac = 0). If the first sample produces 3 or more nonconformities, the lot is rejected with no further samples. If, on the other hand, the first sample produces either 1 or 2 nonconforming items, a second sample of 500 items is required. Then, having taken a second sample, if there are fewer than 3 nonconforming items from both samples, the lot is accepted. If there are 4 or more nonconforming items from both samples, the lot is rejected.

Similarly, in a multiple sampling plan, second, third, fourth, or more samples give the basis for a decision. It is possible this decision arrives after the second sample; however, it may continue through 100% lot inspection. If management feels comfortable with fewer samples, because the sample size is smaller in multiple sampling, they can save a considerable amount of money depending on the expense of inspection.

Other plans include the sequential sampling plan, chain sampling inspection, skip lot, and variable plans. The first two are useful for costly inspection or destructive type inspection whereas skip lot is typically used in laboratory analysis such as chemicals. There are also acceptance plans that require variable characteristic inspection rather than attribute. Each have their main purpose with the common purpose of reducing costs, inspection dangers, and time while maintaining the assurance the lot is acceptable. The majority of acceptance sampling is accomplished with the standard inspection plans already mentioned. This text will not delve into the details of some of these more specialized plans.

## Producer Interests versus Consumer Interests

When a lot arrives in the plant, the customer's main interest is to not accept a bad lot — this is even more important than not rejecting a good lot. From a customer's most strict point of view, it is immaterial if a good lot is rejected. In that case, the worst that happens is that a good lot is not used in the production of a product. (Producers are more concerned about the customer rejecting a good lot.) From the producer's point of view, it is immaterial if a bad lot is accepted — to an extent. Ultimately, the existence of a bad lot hurts the producer's reputation. Thus, between customers and producers, there is a relation that is not necessarily in conflict, but whose interests simply differ from the other. The customer and producer must iron out these differing interests, given in terms of risk probabilities, when they agree on a plan of acceptance.

# 148   Chapter Eight

This relationship is the basis for creating specific acceptance sampling plans between those supplying the lots (producer) and those using the lots (customer). These plans take on characteristics that can be displayed visually by graphing curves based on probabilities associated with both the producer's risk and the customer's risk. Operation Characteristic (OC) curves show at one end the producer's risk (alpha) and, at the other end, the customer's risk (beta). The shape of these OC curves depends on three factors: the population (N), the sample size (n), and the count of nonconformities (c).

## The OC Curve for a Specified Single Sampling Plan

The Operating Characteristic (OC) curve provides an excellent visual means of understanding and analyzing an acceptance plan (Figure 8.2). This curve shows the relationship between the probabilities of acceptance given a certain percent nonconforming. The percent nonconforming is a range of values close to the expected value, where the expected value is based on historic data and the capabilities of the process. For example, over several years of normal operations, a producer found that the percentage of nonconforming product was 1%. This 1% would be a basis for the producer determining a risk they could agree on.

The horizontal axis of the OC curve normally starts at 0 and extends to a percentage of defects past the consumer's risk. Each point on the line represents a probability

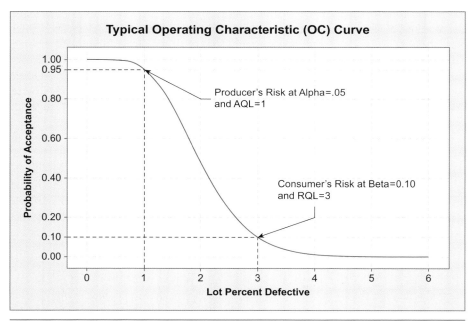

FIGURE 8.2   Typical OC curve, goes around here.

Acceptance Sampling and Inspection  149

the lot will be accepted (vertical axis) if the percent defective is a certain value on the horizontal axis. This line is based on the binomial distribution, but can be approximated by the Poisson distribution as the lot size increases. As far as the customer is concerned, as the percent nonconforming increases, the probability of accepting a bad lot reduces. The lower the percent nonconforming, the higher the probability of accepting a good lot occurs. Conversely, from a producer's perspective, as the percent nonconforming decreases, the less likely it is the customer will reject a good lot. As the nonconforming percent increases, the probability of the customer accepting a bad lot approaches the risk the customer is willing to take.

From the producer's interest, the single sampling plan OC curve is typically based on the statistic called the Acceptable Quality Level (AQL). The AQL specifies the risk of the customer rejecting a good lot when the expected value of nonconforming items is at the known acceptable level. The producer's risk (alpha) is commonly set to 5% — if the quality level from the process is at the AQL, there is a 5% chance of the customer rejecting a good lot. This level is considered an acceptable risk by the producer. In other words, there is a 95% likelihood the customer will accept a good lot. The customer's risk (beta) is set at the limit of where the customer feels comfortable a bad lot may be accepted. This is usually around 0.10 and is based on another statistic called the Rejectable Quality Level (RQL), as used in Figure 8.2. The RQL is that point of incoming quality from the producer that exists at the customer's risk. It limits the likelihood of a bad lot being accepted by the customer. Other acronyms used for the RQL include the Lot Tolerance Percent Defective (LTPD) and the Limiting Quality Level (LQL).

Suppose the aircraft cable customer and the producer want to enter a more organized acceptance agreement that takes into consideration both of their interests. Given $N = 5000$, $n = 200$, $c = 4$, and the expected value of quality from the producer being 1%, we can calculate the probability of acceptance with quality varying from the expected value, and plot the OC curve. Assuming the AQL is set to the expected value, the OC curve can show and reveal values of the producer's and customer's risks. The first step is tabulating np and finding the probabilities needed to create the OC curve as shown in Table 8.6. Follow the steps below to find these values in Appendix A.

Start with proportion and sample size. For the first value, $0.01 \times 200$ gives an $np = 2$. This is the number of nonconforming units that are present in the sample if the

TABLE 8.6  Poisson Values for Various np with c = 4

| p | np(λ) | Pa |
|---|---|---|
| 0.01 | 2 | 0.947 |
| 0.02 | 4 | 0.629 |
| 0.03 | 6 | 0.285 |
| 0.04 | 8 | 0.100 |
| 0.05 | 10 | 0.029 |

expected value of quality is 1%. In Appendix A, using np = 2 (indicated by lambda λ), and a count of 4 (indicated by k), the Poisson distribution gives the value of 0.9473. This is the probability there are 4 or fewer nonconforming units out of 200 when the quality level is 1% nonconforming. Do the same for each of the other proportions of 0.02, 0.03, 0.04, and so on. They are, in respective order, 0.6288 for np = 4, 0.2851 for np = 6, 0.0996 for np = 8, and 0.0293 for np = 10. Past np = 10, the probability is so low it is not necessary to continue for the purpose of plotting the OC curve. Each of these values corresponds to a particular percent nonconforming from the producer on the horizontal axis.

The vertical axis on the OC curve represents this distribution and, as such, goes from 0.00 (no possibility of occurrence) to 1.00 (certainty of occurrence). By placing the percent nonconforming on the horizontal axis (i.e., 0, 1, 2, 3, 4, 5 ...) points are coordinated with the values above to finish the OC curve (Figure 8.3). Alpha is found by following the expected value (AQL) of 0.1 (or 10%) on the horizontal axis vertically up to the curve. Following a horizontal line from that point over to the vertical axis reads 0.95. Regarding beta, the customer's risk, find 0.10 on the vertical axis, follow a horizontal line over to the curve, and then move down vertically to the horizontal axis. There find 4 on the horizontal axis, which represents RQL or 4% nonconforming from the producer. At that point, out of 200 (n), the inspection team may expect 8 nonconforming items.

The reciprocals of these curve probabilities represent the respective risk: either the producer's risk or the customer's risk. For example, the probability for accepting

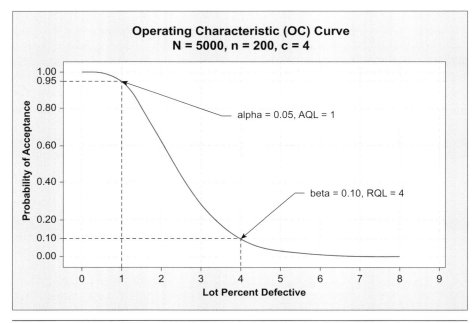

FIGURE 8.3  OC curve for n = 200 and c = 4

a lot when the quality level is 1 percent (same as AQL) is 0.9473. The reciprocal is 0.053 (alpha). The probability for accepting a lot when the quality level is 4 percent (same as RQL) is 0.1 (beta). If the team wants to stipulate alpha as 0.05 rather than the calculated value of 0.053, the corresponding number nonconforming is less than a whole number (approximately 1.98) and the AQL is slightly less than 0.10. If the team (in this case the producer) wants to maintain alpha at 0.05 and the AQL is 0.10, then n has to be different than the original 200. To find n under these conditions, where c = 4:

$$n = np_{(a=0.05)}/P_{(aql)} \quad (8.1)$$

$$n = 1.98/.01 = 198$$

Similarly, plans can favor the customer by stipulating beta and RQL. Keeping the RQL as 4, then find in Appendix A np = 8 (from the example) with c = 4. Because of rounding, the value of 0.10 corresponds directly to the stipulated beta of 0.10. Therefore, this plan can continue with n = 200.

## OC Curve Properties

These plans and subsequent OC curves are representative of a probability distribution, and there is a distinction between a continuous source of product and an infinite source. The distributions and OC curves created so far in this discussion have followed the binomial distribution and have been approximated by the Poisson distribution. These are referred to as Type B curves. With a finite source (small N) of product, these curves are better approximated with the hypergeometric distribution and are called Type A curves. In some cases, the difference can be insignificant; however, Type A curves tend to be steeper (hence, more conservative) in regard to the percent nonconforming against the probability of acceptance.

In practice, the sample size is determined by several methods. Each has its effect on the plan, which is revealed in the OC curve. The lot size N actually has the least effect on the OC curve if the N is high enough to assume normality (type B curve). If the N becomes too small to be approximated by the Poisson distribution, it has a slight effect (type A curve). As discussed above, sample sizes are frequently determined by inspection plan tables where lot size makes less difference as N increases. However, it is also common in practice to determine the sample size as a set percentage of the lot size. This practice leads to a varied OC curve where higher lot sizes create steeper curves. Also, this practice may waste resources when the lot size becomes very large.

Another common practice is setting the sample size n regardless of lot size. In type A curves, this practice results in slight variation among OC curves with varying lot sizes. In this regard, the smaller the lot size is with a fixed sample size, the steeper the OC curve becomes. Similarly, in type B curves, as n increases with a fixed lot size, the OC curve becomes steeper. A steeper OC curve is a more discriminating curve. It reduces the chance of the customer accepting a bad lot and of a producer having a good lot rejected.

The acceptance number also makes a difference. Keeping both the lot size N and the sample size n constant, the OC curve will become steeper as the acceptance number c decreases. Therefore, as c decreases, the plan again becomes more discriminating, hence offering a greater protection from both the producer's and customer's perspective.

## Other Common Acceptance Statistics

The OC curves discussed to this point are representative of the AQL and the RQL statistics. Other statistics produce different curves and focus on specific aspects of the inspection process. What follows is a short description of some of the more commonly used statistics that work in conjunction with AQL and RQL.

### AOQ Curves

If a lot is rejected, the normal procedure is to rectify the lot. To do this requires 100% inspection. The Average Incoming Quality (AIQ) statistic is the average level of quality prior to any inspection and, hence, rectification of a lot. An inspection without removal of any nonconforming units reveals the AIQ. As the lot is inspected, the nonconforming items typically are removed and replaced with conforming items. The resulting rectified lot represents a perfect lot of outgoing quality. However, other lots that are not rejected may have similar properties of the rejected lots. The Average Outgoing Quality (AOQ) is the average quality of the combined lots; it is the basis for the AOQ Curve. This curve fits the AOQ (vertical axis) against the incoming quality (horizontal axis) and shows the probability of nonconforming items entering the manufacturing process. The top of the resulting curve represent the maximum average of nonconforming product. This statistics is called the Average Outgoing Quality Limit (AOQL). Figure 8.4 shows the AOQ curve for the same plan described above, with N = 5000, n = 200, and c = 4. The AOQ is:

$$AOQ = Pap(N-n)/N \tag{8.2}$$

where Pa (estimated by the Poisson distribution) is the probability of accepting a lot given the percent nonconforming (p) and the sample size (n). N is the lot size.

### ATI Curves

The Average Total Inspection (ATI) statistic shows the average number of items inspected given N, n, and c. Because the ATI is an average, it is calculated over a period of time through several lots (Figure 8.5). Although it does not indicate how effective a plan is, it can show how many items the inspection team (including both customer and producer) will expect to examine. As the percent nonconforming increases, the number (on the vertical axis) approaches N (in this case 5000). The ATI is:

# Acceptance Sampling and Inspection 153

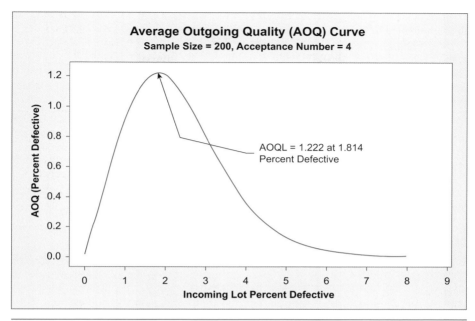

FIGURE 8.4  AOQ curve for N = 5000, n = 200, and c = 4

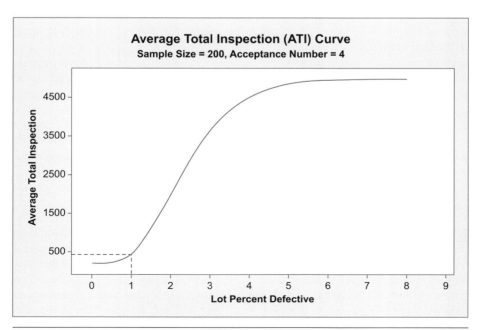

FIGURE 8.5  ATI curve for N = 5000, n = 200, and c = 4

$$ATI = n + (1 - Pa)(n - N) \tag{8.3}$$

where Pa (estimated by the Poisson distribution) is the probability of accepting a lot given the percent nonconforming and the sample size n. N is the lot size. The curve is created by calculating the ATI for each percent nonconforming (the horizontal axis). Note the ATI at the AQL of 1 percent. It is somewhat higher than the n of 200. This is because over time lots will be rejected.

ATI curves are also useful in double and multiple sampling plans. As the reader can imagine, inspecting fewer items becomes more cost effective. The ATI curve easily reveals this information and can serve as a valuable tool in determining the appropriate plan.

## Automated 100% Inspection

The rationale behind these acceptance sampling plans is cost. With automated 100% inspection, both the producer's and the customer's interests are satisfied, with no risk on either side, but the cost is prohibitive under traditional circumstances. The costs involved include labor, upkeep of the instrumentation, and time. The latter may result in loss of production. Some operations have no choice but to do 100% inspection. Typically these include pharmaceutical companies, aircraft components, and others where the risk of failure could cause injury or death. There is clear advantage in these cases to automate. Some operations are forced to do acceptance sampling — for example, any company that conducts exclusively destructive tests in making acceptance decisions. Most operations, however, can take advantage of new technology to move away from acceptance sampling.

Companies that can perform automated 100% inspection should investigate the technology available and make a cost analysis comparing acceptance sampling to implementation of a new automated system. These automated systems can rapidly perform 100% inspection of incoming product for the customer or outgoing product for the producer. It is not uncommon to see producers investing in these systems, accommodating customer demands for quality assurance. Although this may seem as though it is burdensome to the producer, the payback is typically substantial in the resulting decrease in rejected lots from customers, and bolstered marketability.

The technology found in automated 100% inspection remains diverse; it includes vision systems, laser systems, scent detectors, X-Rays, magnets, and a host of others. Some technology for automated 100% inspection can be as simple as a weight scale. Recent improvements, however, have centered on vision systems (Figure 8.6). These systems operate on geometric properties checked against geometric dimensioning and tolerance (GDT), surface dimensions, and other established specifications. They also look for cracks and other non-conformities. A higher resolution capability of the vision system directly relates to a more accurate measure. As this technology improves and becomes more popular, the costs are reduced. Combined with robotics, they can inspect and measure at multiple angles and vantages. Among the advantages are ease of data

## Acceptance Sampling and Inspection 155

FIGURE 8.6 Vision system. *Photo provided by Banner Engineering Corp., www.bannerengineering.com*

collection and ease of data analysis. Automated 100% inspection systems can accommodate both attribute measures and more complicated cases of variable measures

Automated 100% inspection is not without error, but the error is significantly reduced in most cases. Both type I and type II errors can still occur, and rectification ultimately continues. But in many cases, companies are moving away from the traditional acceptance sampling developed in the 1940s for the more sophisticated automated 100% inspection.

## Chapter Eight Discussion

1. What is the advantage of an acceptance sampling plan over 100% inspection?
2. How does stratification control randomness bias? Think of some other ways to control these important study assumptions.
3. Who created the first acceptance sampling plans and how were they used? Which national standard controls the use of these plans?
4. What is an AQL (Acceptable Quality Level) and how does it determine which acceptance plan to use? How is the proper sample size, and rejection limit determined?
5. Explain when to use normal, tightened, or reduced inspection levels.
6. What is the difference between single, double, and multiple sampling plans?
7. Explain the difference between the consumer's interests and the producer's interests.
8. What is an OC curve?
9. What are some other common acceptance statistics used besides the AQL?
10. When is 100% inspection a viable alternative?

## Chapter Eight Problems

1. Select a few items from around the house, school, or workplace, and determine the quality characteristic. Then speculate on how to collect data for acceptance sampling purposes and determine the best way to control randomness and bias.
2. Using a general inspection level of II, a lot size of 720, and AQL of 1.0, what are the acceptance and rejection parameters for a normal inspection, single sampling plan? Which standard was used?
3. Suppose lots in the plan in Problem 2 were rejected several times. How would this differ from the Problem 2 solution?
4. Suppose lots in the plan in Problem 2 were accepted several times. How would this differ from the Problem 2 solution?

# NINE
## Basic Inferential Applications

Thus far, this text has described distributions through a variety of statistics and standard applications in industrial settings. These descriptive techniques and practices culminate in decision making; they suggest trends and patterns, or indicate judgment of a distribution. However, they still remain descriptive in the sense that no confidence in the distribution was established.

This confidence — bolstered by the size of the population and sample, assumptions regarding normalcy, and sampling methods — facilitates decisions going beyond that of purely description. In other words, confidence determines how much statisticians can make inferences regarding a certain distribution. Inferences include making generalizations, conclusions, and decisions toward the population based on a sample of that population. They are also affected by what is known of the population, for example, whether $\mu$ (mu: the population average) or $\sigma$ (sigma: the population standard deviation) is known or unknown. In industry, the overwhelming method for increasing knowledge about the population is to increase the sample size.

In industry, inferential statistics are utilized less than descriptive statistics. The majority of quality activity is done with topics included thus far in this text. However, some research and development questions, as well as operations questions, are best answered with inferential statistics. This chapter will point to some of the more common inferential statistics found in industry. These include confidence intervals and hypothesis testing. Normal distributions will be discussed first followed by inferences regarding distributions that cannot be assumed normal. This chapter will briefly explain concepts such as population versus sample, confidence intervals, hypotheses, and experimental tests. It will utilize statistics and distributions such as the z distribution, z score, the t distribution, t test, $\mu$, $\sigma$, and basic analysis of variance (ANOVA) using the F ratio.

## Confidence Intervals

The most basic aspect of inferential statistics is the confidence interval (CI). The requisite to inference is knowing the population. Without measuring the entire population, this knowledge comes from measuring a sample. But how close is this knowledge to reality? Confidence intervals help establish such information.

Theoretically, if a sample is perfect and fully represents the full population, the estimated values will equal the population values. Only because there is always a difference, however slight, is there the need to establish confidence in how close the sample value is to the real value (population value). The closer the sample is to the population, the higher the confidence is with sample data. The most common application of a CI is to estimate the mean, or μ, of the population. This section estimates an interval (CI) in which the value (μ) exists.

Confidence is simply a probability that becomes known as the confidence coefficient. Common confidence coefficients include 0.90, 0.95, 0.98, and 0.99. But the terminology used in industry stresses percentages rather than coefficients. Therefore, the 0.95 confidence coefficient provides a 95% confidence level. A 0.99 coefficient provides a 99% confidence level.

Given a confidence level of 95%, a confidence interval is taken around the mean of the sample distribution. In a normal distribution, the mean is generally around the middle of the distribution. We can be 95% confident the population mean is within this interval. The remaining 5% is divided with one half on the positive side and the other half on the negative side. There is a 0.025 (2.5% chance) probability the population mean is outside the confidence interval on the high side, and a 0.25 (2.5% chance) probability the population mean is outside the confidence interval on the low side. This combined 5% area is commonly referred to as the risk, denoted by α (alpha).

Recalling the discussion in Chapter Four regarding the Empirical Rule and the Z distribution, normal distributions accommodate approximately 95% of data between 2 standard deviations from the mean. Looking at half the curve for simplicity (Appendix B), find the probability (area) equal to half the distribution. This is 95/2 = 47.5 or 0.4750 in the table. Taking the confidence interval out of the middle of the distribution places the z score at 1.96 standard deviations from the mean. Having 95% confidence in a specific sample requires finding the value that corresponds to 1.96 above and below the mean — hence, the interval.

Here is a short example. A plant manufactures graphite used in Electro Discharge Machining (EDM) applications. To estimate the population parameter μ, they want to determine a 90% confidence interval regarding the average hardness (using the Rockwell L scale, or HRL). After taking 50 randomly selected measurements from the supply of graphite, they find a sample average of 80HRL with a standard deviation of 3.00. Because the sample size n is 50 and considered normal, it is assumed the standard deviation equals σ (sigma). The interval is determined with the following equation:

$$\bar{x} \pm z_{\alpha/2}(s/\sqrt{n}) \tag{9.1}$$

where z for two tails (half the α) is found in Appendix B. 90/2 = 45; hence the area value is 0.450. Because there are only values of 0.4505 and 0.4495, and 0.450 is in the middle of those two values, add the next level of accuracy as 0.005 making z = 1.645. So:

$$80 \pm 1.645(3/7.071)$$
$$80 \pm 0.7$$

To interpret this, the graphite plant could consider the population average μ existing within an interval between 79.3HRL and 80.7HRL with a confidence of 90% (Figure 9.1).

## Confidence Intervals with Unknown σ

The estimations in this example assume knowledge of σ because the sample size n is high, and there is prior knowledge the distribution is normal. If this normality assumption is not forthcoming, methods for estimating CIs are based on more conservative distributions. The smaller the n, or for whatever reason the distribution is not considered normal, the more conservative the distribution becomes in making inferences about the population. In other words, if the distribution is normal, the confidence levels are approximated against the empirical rule. As the distribution becomes less normal, typically because the n is small, what was a 90% area represented in a normal distribution must now accommodate a wider area to maintain the required confidence.

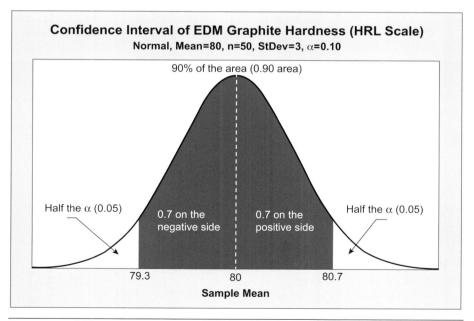

FIGURE 9.1   90% CI estimating μ for EDM graphite hardness

A century ago, William Gossett was presented with such a dilemma during his tenure at Guinness Breweries in Dublin. Because brew sample sizes were small, tests using the normal distribution delivered erroneous results over a period of time. To remedy this, he developed a set of distributions called the *t distribution* with widened tails and flatter kurtosis. As the n diminished, these characteristics increased; he described each of these distributions with what he called a *degree of freedom* or n − 1. As n increased, the distributions approached that of a normal distribution.

In very simple terms, the degree of freedom reduces the sample size to account for the sample average having already been calculated. For those distributions where n is high, n − 1 makes little difference. In fact, a comparison between the t distribution (Appendix C) and the normal distribution (Appendix B) will reveal the same areas when the degrees of freedom are at infinity. For those distributions where n is low, the effect of subtracting one from n makes a distinct difference. For example, if n = 5, n − 1 leads to a 20% reduction of data. This result widens tails and flattens kurtosis in the sample distributions and requires further estimates to be conservative. The t distribution indicates that as the degrees of freedom decrease, the higher the t score becomes. In other words, the t distribution with lower degrees of freedom must reach farther with more standard deviations from the mean to maintain the same level of confidence.

Imagine taking a random sample — perhaps only a drop — of stout (dark beer) from a finished brew, inactivating the yeast with a dilution of mercuric acid, and then agitating the sample to disperse the yeast cells evenly over the entire volume of the liquid sample. While the yeast cells are still suspended evenly, the sample is placed on a hemocytometer (a microscope slide with a counting chamber attached where the cover slip would normally be). A solidifying agent gels the liquid sample to freeze the yeast cells in place.

This particular hemocytometer sections one square millimeter into 400 visible zones. The yeast cells are counted zone by zone through a microscope, results are recorded (Table 9.1), and the process is repeated with another sample of stout. After repeating this for several more drops of stout (in this example, 18 drops), more data

TABLE 9.1    Count of Yeast Cells

| 6 | 1 | 5 | 1 | 5 | 8 | 7 |
|---|---|---|---|---|---|---|
| 7 | 5 | 8 | 8 | 2 | 7 | 8 |
| 4 | 8 | 7 | 6 | 4 | 6 | 9 |
| 4 | 6 | 1 | 3 | 1 | 12 | 4 |
| 5 | 11 | 6 | 12 | 12 | 4 | 4 |
| 9 | 11 | 11 | 12 | 4 | 12 | 10 |
| 9 | 12 | 4 | 9 | 4 | 2 | 5 |
| 2 | 2 | 5 | 6 | 8 | 1 | 3 |
| 9 | 4 | 5 | 1 | 8 | 7 | 10 |
| 5 | 5 | 2 | 1 | 3 | 2 | 10 |
| 9 | 11 | 11 | 5 | 12 | 6 | 10 |

TABLE 9.2   18 Averages of Yeast Count per Drop of Stout

| 6.2425 | 6.595  | 6.3825 | 6.54  | 6.76  | 6.81   |
|--------|--------|--------|-------|-------|--------|
| 6.5525 | 6.4425 | 6.3925 | 6.425 | 6.375 | 6.5875 |
| 6.4975 | 6.6875 | 6.3975 | 6.335 | 6.8   | 6.635  |

are recorded to determine a confidence interval estimating the $\mu$ for the entire batch of stout. There are several reasons more data are not collected. This process is tedious, labor intensive, and constrained by time before the batch yeast cell count changes. All 18 data, where each datum represents the average number of yeast cells within a drop of stout, are represented in Table 9.2 (see Appendix E). Computing the sample distribution statistics yields a mean of 6.53 and a standard deviation of 0.166.

This particular chemist wishes to develop a 99% confidence interval to infer the value of the population mean of the entire batch of stout. Why so high? The yeast content of the stout, among other factors, can indicate the consistency of the brewing process. If the process is consistent, the brewery can maintain a certain level of yeast and thence quality. Given the delicate process of brewing beer, the brewery and the chemist must be as close to certain regarding yeast content as allowed. Equation 9.2 helps calculate the CI, using the t distribution. Remember, n=18; therefore, the degrees of freedom equal 17 (df = n − 1).

$$\bar{x} \pm t_{(n-1)(\alpha/2)} (s/\sqrt{n}) \tag{9.2}$$

where $\alpha$ is the upper critical point based on the percent of confidence.

Looking at Appendix C, locate the row under the df column corresponding to 17 (df = n − 1). Requiring 99% confidence means having an $\alpha$ of 0.01. Because this confidence is taken out of the middle of the distribution, $\alpha$ (risk) is divided with half on the positive side of the curve and the other on the negative side (Figure 9.2). The critical value is (1 − .005) from the top, or found using the column for $\alpha = 0.005$. This gives the value of 2.898. With the degrees of freedom, and the level of confidence, the critical value is 2.898 standard deviations from the mean in this conservative distribution.

To continue with the yeast data using equation 9.2, the 99% confidence interval containing $\mu$ is:

$$6.53 \pm 2.898\,(.166.166/\sqrt{18})$$
$$6.53 \pm 2.898\,(.166/4.242)$$
$$6.53 \pm 2.898\,(.039)$$
$$6.53 \pm .113$$

The stout chemist can be 99% confident the $\mu$ (population average of yeast levels) exists somewhere between 6.417 and 6.643. The brewery can now decide (presumably based on taste) if the process needs to increase, decrease, or maintain current levels of yeast in the stout to control quality.

Gossett published his research regarding the t-distribution in *The Probable Error of a Mean*. Although he published this research with permission from the Guinness

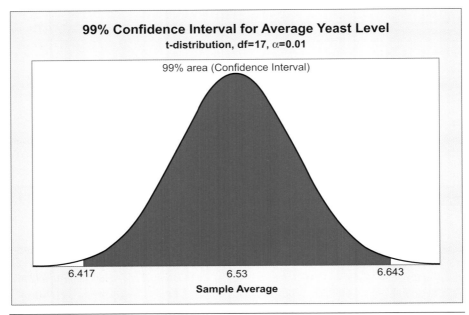

FIGURE 9.2   Distribution showing 99% CI for average stout yeast content

Brewery, they required he use a pseudonym. His was "Student," leading to the accepted term for this distribution as "student's t."

## Experimental Tests

Although inferential, confidence intervals do not test a specific point. When defining points, or comparing scores, we wish to make an inference to whether those values are what they purport to be, or perhaps if they are simply different within significant terms. To make any such test requires a hypothesis, which is a proposed construct asserting an outcome — for example, "I think basketball players around here are taller than 6 feet." Previous studies have claimed the height of basketball players in this location to be average of 6 feet tall. Although this hypothesis is crudely stated, and not stated in scientific terms, it presents a measure for basketball player heights that is believed to exist at the specific location, a measure that is conflict with the established value. To fine tune the hypothesis into scientific terms, the first step is to remove "I think." Removing "I think" changes the statement from an opinion to an assertion. The data become heights of basketball players, and the statistics include the mean and standard deviation of the sample. "Basketball players around here are taller than 6' feet," tests only whether or not the mean is greater than 6 feet. It makes no assertions regarding equal to, less than, equal or less than, or equal or greater than. The only way to accept the revised hypothesis is if the mean of the basketball players in that area *is greater than* 6 feet within a level of confidence — or rather, to a level of significance.

The next step is writing the hypothesis in scientific terms using the appropriate symbols. $H_o$ refers to the null hypothesis, which represents the conventional or accepted status quo in mathematical terms. For this example, the null hypothesis is

$$H_o: \mu = 6'.$$

$H_a$ represents the research hypothesis; it is sometimes referred to as the alternate hypothesis. It is not necessarily the opposite of the null hypothesis (i.e., $H_a: \mu \neq 6'$), but always tests a conflicting assertion against the accepted $H_o$ value. In this case, the research hypothesis is

$$H_a: \mu > 6'$$

The experiment tests this hypothesis with a value based on sample distribution statistics, size of the sample, and how normal the sample distribution is. In this case, the test is based on the sample mean ($\bar{x}$) and becomes the statistic tested. However, this mean must be compared to the $\mu$ and any difference given as the number of standard deviations away from $\mu$. So, the test statistic (T.S.) is actually the z score (assuming a normal distribution) computed by:

$$z = (\bar{x} - \mu) / (\sigma/\sqrt{n}) \tag{9.3}$$

The T.S. is then held against a critical value (CV) that marks the beginning of a rejection region (R.R.) found under the area of the distribution. If the distribution is normal, the R.R. is found using a z score (Appendix B). If the distribution is not normal, or where n is low, the R.R. is found using a t score (Appendix C). If the T.S. falls into the rejection region, the null hypothesis is rejected and the alternate hypothesis is retained. In this particular case, the R.R. amounts to the area above the CV, depending on a level of confidence. The CV is lower given a 90% confidence (or risk of $\alpha = 0.10$) or higher given a 98% confidence (or risk of $\alpha = 0.02$). Given a certain level of confidence, if the basketball players in this area are indeed taller than 6', data will be in or above that rejection region of the distribution that has a mean of 6' (72 inches).

To calculate the test and finish the example, suppose the heights of 50 basketball players in the area of question were measured and the descriptive results were a mean of 6' 1" (73 inches) with a standard deviation of 3". Those involved realize this is a normal distribution and wish to maintain a risk of $\alpha = 0.05$ (95% confidence). First, to give the example formal notation and to represent everything in inches:

$$H_o: \mu = 72$$
$$H_a: \mu > 72$$
$$\text{T.S.: } z_{\bar{x}} \text{ (with } \bar{x} = 73\text{)}$$
$$\text{R.R.: with } \alpha = .05$$

given a one-tail test, the area under the normal curve at 95% gives a CV of 1.645 (from Appendix B).

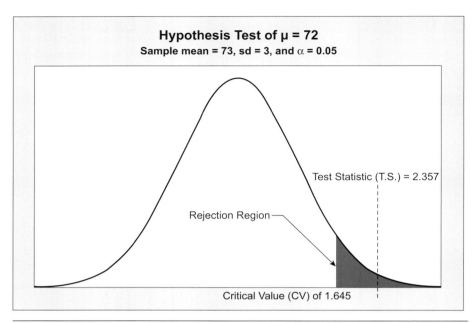

FIGURE 9.3   One-tail test rejecting Ho: μ = 72 and retaining Ha: μ > 72

To compute the T.S.:

$$z = (\bar{x} - \mu) / (\sigma/\sqrt{n})$$
$$z = (73 - 72) / (3/\sqrt{50})$$
$$z = 1 / .424268 = 2.357$$

As the T.S. (z score of $\bar{x}$) resides well into the R.R. beyond the CV of 1.645 (Figure 9.3), the null hypothesis is rejected and the research hypothesis is retained with a 95% level of confidence. Thus, it is true (with 95% confidence) the "basketball players around here are taller."

## One Tail versus Two Tails

Because the $H_a$ in this example is concerned only with $\bar{x}$ being greater than 6', this test is called a one-tail test. The R.R. is at the top of the curve and the risk is placed entirely on one side. The 0.95 confidence coefficient (1 − α) comes from a risk (α) of 0.05. Because it is a one-tail test, the risk calls for the placement of the R.R. at an area equal to 95% of the normal curve. Glancing through the area values of that curve (Appendix B) we can find 0.9495 and 0.9505 in the very next column. The z score for 0.9495 is 1.64. Adding 0.005 to this z score gives a close approximation of 0.95.

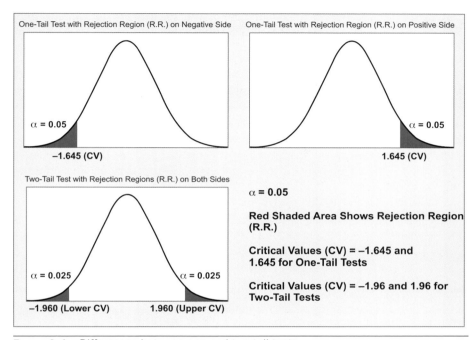

FIGURE 9.4    Differences between one and two tail tests

Had the hypothesis asserted the basketball players were shorter, with everything else remaining the same, the R.R. would have been on the negative side of the curve (Figure 9.4, upper left). With an $\alpha$ of 0.05, the R.R would start at the negative z score of $-1.645$ (CV). Determining rejection or retention of a hypothesis proves to be one of the most common confusions regarding hypothesis testing. One rule of thumb, given a one-tail test, is to remember the R.R. is always on the same side as the $H_a$.

Another version of the $H_a$ would hypothesize the basketball players were not equal to $\mu$. This would be a two-tail test (Figure 9.4, bottom). Being not equal can be either above or below the mean. Therefore, this scenario divides the risk in two ($\alpha = .05/2$) While maintaining a risk of $\alpha = 0.05$, half of the alpha is on the positive side of the curve and half is on the negative side. The area under the normal distribution (Appendix B) that marks the beginning of the R.R. is 0.9750 on the positive side and 0.0250 on the negative side. This corresponds to z scores of 1.96 on the positive side and $-1.96$ on the negative side.

## Probability of Error in Hypothesis Testing

As with any statistical test, there is the probability of error. In inferential statistics, including hypothesis testing, two types of errors are commonly defined as *Type I* errors and *Type II* errors.

A Type I error occurs when the null hypothesis is rejected but, in reality, is true. This occurs with a probability of $\alpha$ and in industry is frequently called a *false negative*. A Type II error occurs when the null hypothesis is retained but, in reality, is false. This occurs with a probability of $\beta$ and in industry is frequently called a *false positive*. Although $\alpha$ is relatively simple to understand and compute, $\beta$ is not. Ideally, you would want to reduce both. But there is a direct tradeoff between these two errors. As $\alpha$ decreases, $\beta$ increases. Power is the ability to decide correctly. Because power = $1 - \beta$, as $\beta$ increases, power decreases. In hypothesis testing, $\beta$ decreases (power increases) as the difference between the mean and $\mu$ increases. Likewise, as this difference decreases, the probability of committing a Type II error increases. Those studies where the mean and $\mu$ are very close and $\alpha$ is very low make for the worst case scenarios.

This difference, and the described $\beta$, may subsequently have an effect on researchers rejecting the null hypothesis or simply remaining inconclusive. $\beta$ varies for each possible population $\mu$, but can be considered for specific values. Given that, $\beta$ is a probability and is described by the following:

$$\beta = P \{z < z_a - [(\bar{x} - \mu) / (\sigma / \sqrt{n})]\} \tag{9.4}$$

where

$z_a$ = the z of the research hypothesis or critical value (CV)

$(\bar{x} - \mu) / (\sigma / \sqrt{n})$ is the same as the test statistic (T.S.)

P = the probability under the normal curve

Note this example is one tail and based on the z distribution. The formula for $\beta$ differs slightly based on other scenarios. Consider a test where

$H_o$: $\mu = 200$
$H_a$: $\mu > 200$ where n = 50, $\bar{x} = 205$, and s = 22

The type I risk $\alpha$ is set to 0.05. The T.S. is

$$z = (\bar{x} - \mu) / (\sigma / \sqrt{n}) = 1.607$$

The R.R. starts on the positive side of the curve at the CV, 1.645 (Appendix B). Even though the T.S. does not enter into the R.R., conservative statisticians would not necessarily retain the null hypothesis. Nor would they reject the research hypothesis at this point. The result simply remains inconclusive pending further information. Identifying $\beta$ reveals more information. Using equation 9.4:

$$\beta = P \{z < z_a - [(\bar{x} - \mu) / (\sigma / \sqrt{n})]\}$$
$$= P [z < 1.645 - 5/(22/7.071)]$$
$$= P [z < 1.645 - 1.607]$$
$$= P [z < 0.038]$$

With z = 0.038, the corresponding area in the normal distribution (Appendix B) is 0.5152 (with some interlinear interpolation) and defines $\beta$ given $\bar{x} = 205$. Because the researchers are not willing to risk a type II error with a probability this high ($\beta$ = 0.5152), they decide to remain inconclusive. Had $\beta$ been smaller, such as 0.15, for example, they may have been willing to accept the risk and conclude by retaining the null hypothesis. Incidentally, the power of this example is $1 - \beta$ or 0.4848, further supporting an inconclusive decision.

There are methods for minimizing the probability of both types of error. Some of these include choosing one-tail tests when possible, keeping $\alpha$ at a reasonable level, and, as usual, providing as large a sample size as possible.

## Testing $\mu$ With Unknown $\sigma$

Hypothesis testing with unknown $\sigma$ is treated basically the same way as confidence intervals with unknown $\sigma$. The reason $\sigma$ is unknown is generally because of a low n. This low number of samples could be from a slow process or a dangerous environment during sampling, or simply because the product is expensive or testing destroys the sample (destructive testing). Other than the low number of samples available for the test, the design of the experiment is the same as when $\sigma$ is known. Consider the following example.

A supplier claims the annealed hardness of their sheet metal is 55HRC (Rockwell Hardness C Scale). In a subcontracted operation, a small plant is using the sheet metal to fabricate riding lawnmower fenders. However, upon purchasing and using the sheet metal for a short period of time, operators report an unusual increase in die wear and malfunctions in the stamping machines.

Both operators and line supervisors observe a difference in the sound and vibrations while the machine is in operation. Subjectively, those working with these machines assert the difference is due to the sheet metal being harder than what is recommended. This observation is congruent with the reports of unusual die ware. The floor manager, who wants some objective information prior to approaching the purchasing department with this problem, directs a team to test the hypothesis that the hardness of the sheet metal is different than the hardness reported by the supplier. With $\alpha = 0.05$, the experiment is set up as follows:

$$H_o: \mu = 55HRC$$
$$H_a: \mu \neq 55HRC$$

This operation is a small one and the plant orders only one roll at a time. Sample distribution size is limited (n = 10). Data are collected by randomly selecting 10 points along the roll of sheet metal the plant owns. This selection is accomplished by transferring the roll between two large rollers to those selected points, taking a sample, and measuring the hardness with the HRC scale. After measuring hardness, sample calculations give a mean of $\bar{x} = 57$ and a standard deviation of s = 15.

Notice the floor manager's exact wording. She wants to test if the hardness is "different" — not whether the material is harder or softer, but different. Her objectivity configures the study into a two-tail test with the risk ($\alpha = 0.05$) split between the positive and negative ends of the distribution. The area containing rejection regions exists on the negative side from negative infinity to 0.025 of the distribution, and on the positive side from 0.975 to infinity. In a normal distribution, these regions correspond to z scores of −1.96 and 1.96 respectively. However, with n = 10, s of the sample distribution cannot be assumed to equal σ and the sample cannot be assumed normal. Instead, the sample distribution must be compared to the t distribution with a degree of freedom equaling 9 (df = n − 1) rather than the normal distribution.

To finish setting the experiment:

$H_o$: μ = 55HRC
$H_a$: μ ≠ 55HRC
T.S.: t score of the sample distribution mean (57HRC)
R.R.: for a two-tail test with α = 0.05 and df = 9

Critical values on the t distribution end at −2.262 on the negative side of the distribution and begin again at 2.262 on the positive side of the distribution. These values can be found on the standard percentage points of the t distribution (Appendix C).

To compute the T.S:

$$t = (\bar{x} - \mu) / (s / \sqrt{n}) \tag{9.4}$$
$$t = (57 - 55) / (15/\sqrt{10})$$
$$t = 2 / (15/3.1623)$$
$$t = 2 / 4.7434$$
$$t = .4213$$

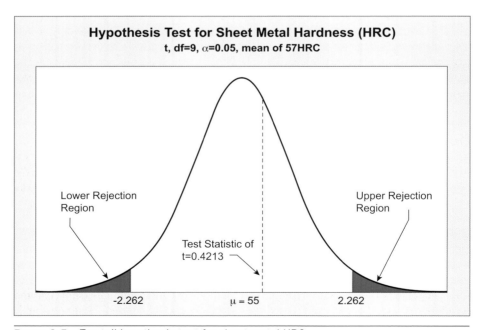

FIGURE 9.5   Two-tail hypothesis test for sheet metal HRC

Because the T.S. is well away from either rejection region and somewhat close to the middle of the curve, it is safe to proceed with retaining the null hypothesis.

The findings indicated the annealed hardness of the sheet metal is actually as the supplier asserted. Although having averted an embarrassing confrontation with the supplier, operations is still perplexed regarding the peculiar behavior of the machines and the unusual die ware. Out of sheer intuition, the line supervisor picks up a scrap piece of sheet metal ejected from the stamping machine and tests the hardness. The reading is 74HRC. Given the only mechanisms between the roll of sheet metal and the stamping machine are the guide rollers, the supervisor begins examining the guide rollers. As it turns out, the guide rollers are too tight and the sheet metal is work hardening prior to reaching the stamping machine. Upon proper adjustment of the process, the problem is solved.

## Tests Between Two Populations

A test between two populations is among the most useful tests available, not only for operations, but also for research and development (R&D). It is less complicated than Analysis of Variance (ANOVA), which uses the F ratio and is covered later in this chapter. It fits within all the discussion covered thus far regarding known and unknown $\sigma$, one or two tails, hypothesis formulation, types of error, and power. The basic difference between two-population tests and tests regarding $\mu$ is comparing one distribution to another distribution rather than an established value (namely the mean). With some modifications regarding the distribution statistics and other minor changes in degrees of freedom when required, the rest of the study is similar.

This test is typically used in operations to determine which of two eligible materials is better for the process. If the distributions in question have a high enough n to reveal central tendency and appear to be normal, $\sigma$ and $\mu$ are considered known. The curve is assumed normal. In this case, the T.S. becomes a z score. If n is low, preventing a known $\sigma$ and $\mu$, the T.S. becomes a t score. (There are variations of this type of test affected by known and unknown variance, independence considerations, equal or unequal variances, and many other factors. This text will give only the most basic of these as a simple example.)

A printed circuit board (PCB) manufacturer uses a conformal coating on the finished PCB to protect the item from contamination, heat, chemicals that cause corrosion, and static discharges. This particular manufacturer sprays the solution on the product and then cures the treated PCB in a heated environment for a specified time. This process creates a bottleneck in the operation, prompting the design engineers to search for an alternative in conformal coatings. They purchase their current solution from a long-time supplier who insists it is the best application. Of primary importance in considering a new solution are a short cure time with little or no additional curing technology and a superior dielectric strength. A new supplier approaches the manufacturer and claims their acrylic-based product will air dry in 15 minutes (short enough to allow continued operations), costs less than their current solution, and withstands 1000 volts with a 1 mil thickness.

The engineers decide to test these claims in a preliminary study. If results are promising, they will conduct a full-scale experiment. As the drying time is relatively simple to confirm, they turn their attention to dielectric strength. The engineers not only want to know if the PCBs can withstand 1000 volts, they also want to see how much they can withstand before failure. This means the PCB used in the test will be destroyed at a considerable expense. The engineers determine an experiment with a small n. They use 14 PCBs in all, with 7 having been sprayed with the current conformal coating and 7 sprayed with the new conformal coating. Great care is taken to make all applications equal and consistent in preparation and in testing. Those involved assume independence between the two samples, with both samples being from a normal population, and equal variance. Statistically, the experiment is configured as follows:

$$H_o: \mu_1 = \mu_2$$
$$H_a: \mu_1 < \mu_2$$

The first sample $\mu_1$ is the PCB with the current conformal coating and is estimated with $\bar{x}_1$. The second sample $\mu_2$ is the PCB with the new conformal coating and is estimated with $\bar{x}_2$. Because the engineers at the plant are not concerned if the dielectric strength of the new conformal coating is the same or lower than the current application, the experiment is configured as a one-tail test. $H_o$ states there is no difference between the current conformal coating and the new conformal coating (status quo). $H_a$ states the current conformal coating has less dielectric strength than the new conformal coating (what the new supplier claims). The data are reported in Table 9.3 (see Appendix E).

The current applications result in a mean of $\bar{x}_1 = 1038$ and a variance of $s_1^2 = 86$. The new applications result in a mean of $\bar{x}_2 = 1059$ and a variance of $s_2^2 = 62$. With the low n, the T.S. is a t value with combined degrees of freedom of $n_1 + n_2 - 2 = 7 + 7 - 2 = 12$. In addition, because each sample has the same n, the sample variances estimating the population variances are assumed equal and are combined to find a pooled standard deviation as follows (Note the variances used in the equation):

$$s_p = \sqrt{[(n1-1)s_1^2 + (n2-1)s_2^2]/(n1+n2-2)} \quad (9.5)$$

$$= \sqrt{[(7-1)86 + (7-1)62]/(7+7-2)}$$

$$= \sqrt{(516+372)/12}$$

$$= \sqrt{888/12} = \sqrt{74} = 8.6$$

TABLE 9.3 Dielectric Strength of Conformal Coating in 1kV/Mil

| $x_1$ | 1028 | 1031 | 1041 | 1052 | 1046 | 1037 | 1028 |
|---|---|---|---|---|---|---|---|
| $x_2$ | 1069 | 1054 | 1063 | 1061 | 1051 | 1067 | 1049 |

## Basic Inferential Applications

The T.S. was computed as follows:

$$\text{T.S.: } t = (\bar{x}_1 - \bar{x}_2) / (s_p \sqrt{1/n1 + 1/n2}) \quad (9.6)$$

$$= -21.57 / (8.6\sqrt{.143 + .143})$$
$$= -21.57 / [8.6(.535)]$$
$$= -21.57 / 4.6$$
$$= -4.69$$

The R.R. is determined using $\alpha = 0.05$. With a one-tail test and the degrees of freedom as df = 12, the area under the t distribution gives a CV of 1.782 (from Appendix C). With a negative CV of $-1.782$, the complete experiment is reported as follows:

$H_o: \mu_1 = \mu_2$
$H_a: \mu_1 < \mu_2$
T.S.: $t = -4.69$, with df = 12 and $\alpha = .05$
R.R.: starts at CV = $-1.782$

We can see immediately that the T.S. of $-4.69$ goes well below the CV of $-1.782$ (Figure 9.6). The null hypothesis of having no difference between the sample means is comfortably rejected in favor of the alternate hypothesis that the new conformal coating has a higher dielectric strength than the current conformal coating. With this, the engineers determine a full scale study is warranted to aid in the decision to abandon

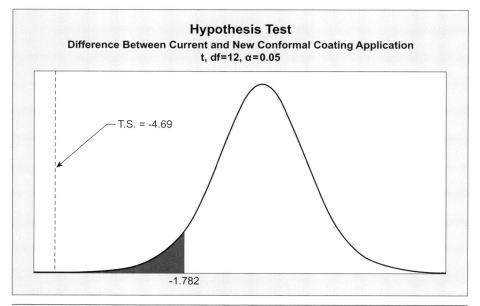

FIGURE 9.6   One-tail hypothesis test for differences in two populations of conformal coating dielectric strength

the current conformal coating and reconfigure the process with the new conformal coating.

Incidentally, the rejection of the null hypothesis is very confident in this test. By glancing through the t distribution for this scenario, we can see it is significant even beyond the 0.0005 α level (99.95% confidence level) evident by Figure 9.6. Using a significance calculator or a *p-value* calculator (commonly available online for free) one can find the confidence level for this test is greater than 99.97%, given a p-value of p < 0.000262. The p-value is the probability the null hypothesis will be rejected in the case it is true (type I error).

## ANOVA

In this example, there are only two populations to compare. What if there are more? Even with only one additional population, there would be three tests to complete. Consider samples 1, 2, and 3. Comparing each sample to each other sample requires 1-2, 1-3, and 2-3, as per the combination equation. If there are five populations, performing multiple t-tests requires

$$C_{n,r} = n! / r!(n-r)! = 10 \text{ combinations} \qquad (3.2)$$

Analysis of Variance, or ANOVA, allows testing multiple populations with one test. This one test determines if there is a significant difference in the means of these populations.

The concept behind this test is analyzing the variance within each sample and comparing that statistic to the mean variance between each sample. This is done through a ratio called the *F ratio*. If the variance within all samples is the same as the mean variance between the samples, the F-ratio is small. If the variance within all the samples differs significantly compared to the mean variance between samples, the F ratio is high. By defining an F distribution with certain parameters, significance can be found using hypothesis testing similar to testing with z and t.

Recall the discussion on subgroups in Chapter Six. Some subgroups can be very close to the centerline ($\bar{\bar{X}}$) and have little variance (close to the LCL on the R chart). Others are close to the centerline but have considerable variance. Perhaps that subgroup is even out of control above the R chart UCL.

Think of the samples in an ANOVA study the same way. If there is very little variance within the samples (close to the centerline, but near the LCL on the R chart), any difference among means will easily be apparent. If there is high variance within the samples (close to the centerline, but near the UCL on the R chart), the means will be surrounded by sample values from each of the samples because they cover a wider area. Now think about comparing three or more of those subgroups or, in this case, samples.

Figure 9.7 shows the difference. Each graph holds three samples — each with respective means of 50, 51, and 52. Yet the distributions in the left graph are distinctly different than the distributions in the right graph. In the left graph, if the variances within the samples are compared to the mean variance between samples, the ratio will

## Basic Inferential Applications 173

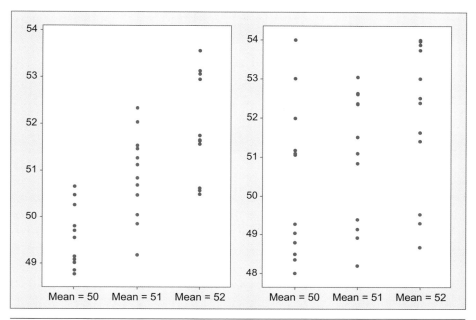

FIGURE 9.7 Comparison of variance with equal means

be high and may be significant. However, for the right graph, the variance within the samples is quite large, as is the mean variance among the samples. Hence, the ratio will be small, indicating no significant difference.

Moving through a short example will help illustrate this. A concrete casting facility is considering the use of plasticizers in their mixture for an upcoming contract to cast sections for a tunnel wall. These sections are intricate; they must adhere to strict guidelines regarding strength, shape, and flow coverage. The plasticizer assures quality in flow coverage. The design team has narrowed down four possible suppliers and will make the decision based on superior flow.

Plasticizer samples are acquired from all four suppliers. Given time constraints, the mixtures must be tested quickly and consistently before each sample cures. Therefore the n is kept low at 8 for each of the four supplier samples for a total of 32. To collect the data, mortar is mixed with the recommended minimum water content, and then filled into a brass flow table mold. The mold is lifted off of the motorized flow table and the table is agitated 25 times for 15 seconds. This step allows the mortar to spread out over the flow table evenly in a circular pattern. After agitation, the diameter of the flowed mortar, which is a function of the plasticizer, is measured in inches. Larger diameters of flowed mortar indicate a better flow. These data are reported in Table 9.4 with generated statistics summarized in Table 9.5 (see Appendix E).

## 174  Chapter Nine

TABLE 9.4  Mortar Diameters on Flow Table in Inches

| S1 | 10.9 | 11   | 11.4 | 11.8 | 11.3 | 11   | 10.2 | 10.8 |
| S2 | 11   | 10   | 11.4 | 10.6 | 11.3 | 11.4 | 11.8 | 10.8 |
| S3 | 11.6 | 12.2 | 11.6 | 11   | 11.9 | 12   | 12.2 | 11   |
| S4 | 11.3 | 11.1 | 11.4 | 11.3 | 11.4 | 11.7 | 11   | 11.6 |

TABLE 9.5  Summary Statistics for Mortar Diameters

| Groups | Count | Sum  | Average | Variance   |
|--------|-------|------|---------|------------|
| S1     | 8     | 88.4 | 11.05   | 0.22285714 |
| S2     | 8     | 88.3 | 11.0375 | 0.31982143 |
| S3     | 8     | 93.5 | 11.6875 | 0.23267857 |
| S4     | 8     | 90.8 | 11.35   | 0.05428571 |

The statistical test is conducted much in the same way the t test was conducted. With $\alpha = 0.05$, the hypotheses and test are configured as follows:

$H_o$: $\mu_1 = \mu_2 = \mu_3 = \mu_4$
$H_a$: At least one mean differs
T.S.: F ratio with $df_b=3$, $df_w=28$, and $\alpha = .05$
R.R.: starts at a CV = 2.947

where $df_b$ (degrees of freedom between groups) is in regard to 4 samples − 1 = 3, and $df_w$ (degrees of freedom within groups) is in regard to 32 items in all minus 1 for each of the four samples = 32 − 4 = 28. Using the information in both the data table (Table 9.4) and the summary table (Table 9.5), calculate what are called the *total sum of squares* (TSS). The two parts of TSS are the *sum of squares between* (SSB) and the *sum of squares within* (SSW). The TSS is computed as follows:

$$TSS = \Sigma x^2 - (GT)^2/n \qquad (9.7)$$

where $\Sigma x^2$ is the summation of all squared measurements and the GT is the grand total of all measurements.

Using Equation 9.7 to continue:

$$TSS = 4080.6 - 130321/32$$
$$= 4080.6 - 4072.53 = 8.07$$

Now for the SSB,

$$SSB = \Sigma t^2/n_s - GT^2/n \qquad (9.8)$$

where $t^2$ is the squared total of each respective sample and $n_s$ is the n of each respective sample (8 in this case). Therefore,

$$SSB = 4074.8 - 4072.53 = 2.26$$

TABLE 9.6  ANOVA Summary Table for Mortar Diameters

| Source of Variation | SS | df | MS | F | P-value | F crit |
|---|---|---|---|---|---|---|
| Between Groups | 2.26125 | 3 | 0.75375 | 3.63409384 | 0.024809 | 2.946685 |
| Within Groups | 5.8075 | 28 | 0.20741071 | | | |
| Total | 8.06875 | 31 | | | | |

and

$$\text{SSW} = \text{TSS} - \text{SSB} \qquad (9.9)$$
$$= 8.07 - 2.26 = 5.81$$

With these values in hand it is time to create what is called an *ANOVA Summary Table* (Table 9.6). This table organizes calculated information and helps determine the F-ratio needed to complete the test.

Note the TSS, SSB, and SSW under the SS column. The df column shows the degrees of freedom as 3 for between groups (4 − 1=3) and 28 for the degrees of freedom within groups (32 − 4=28); the total is 28 + 3=31. The next column shows what is called the *mean square* (MS). The MS between groups is defined as:

$$s_b^2 = \text{SSB}/\text{df}_b \qquad (9.10)$$
$$= 2.26 / 3 = 0.753$$

and

$$s_w^2 = \text{SSW} / \text{df}_w \qquad (9.11)$$
$$= 5.81 / 28 = 0.2075$$

Using the mean squares, the F ratio is calculated as:

$$F = s_b^2 / s_w^2 \qquad (9.12)$$
$$= 0.753 / 0.2075 = 3.63$$

As for the test decision, recall the test was organized as follows:

$H_o$: $\mu_1 = \mu_2 = \mu_3 = \mu_4$
$H_a$: At least one mean differs from the others
T.S.: F ratio with $\text{df}_b = 3$, $\text{df}_w = 28$, and $\alpha = .05$
R.R.: starts at CV = 2.947

At 3.63, the F ratio is into the rejection region beyond the CV of 2.947 (found in Appendix D), and the null hypothesis is rejected in favor of the alternate hypothesis. Figure 9.8 shows the F distribution with the T.S. as stated.

In short, there is a significant difference between at least one of the sample means compared to the others. A look at the means can tell the design team which one, or which ones, were different. Sample 3 has the highest mean. However, Sample 4 has a high mean as well that serves the purpose of the operation and its variance is lower than the rest. This indicates a greater consistency in flow quality of the cement and, hence, a tighter manufacturability of the cement tunnel wall section.

**176** Chapter Nine

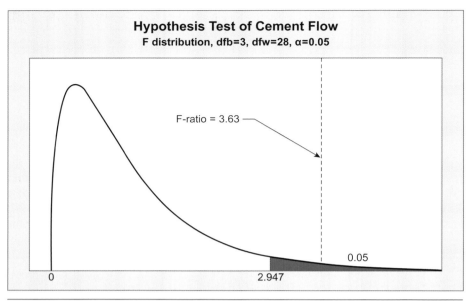

FIGURE 9.8   Hypothesis test of cement flow

Incidentally, the second-to-last column in the ANOVA Summery Table is the p-value. This value represents the probability that a type I error will occur as the null hypothesis is rejected. In this case, it is $p < 0.025$, which means this study was concluded with a confidence level of 97.5%.

## Chapter Nine Discussion

1. What is the difference between descriptive statistics and inferential statistics?
2. What is the distinction between a sample and a population?
3. How are confidence intervals used to infer from sample to population?
4. What is a null hypothesis and what is the difference between it and the alternate hypothesis? What are the other components of a typical hypothesis study?
5. What is the difference between a one-tail and a two-tail test? How can one determine when to reject a hypothesis?
6. What is $\alpha$? What is $\beta$ and what is its relation to the power of the test?
7. When does one use the z test as opposed to using the t test?
8. What is the meaning of unknown variance? How would it affect the study?
9. Describe how to use the t distribution to reject a hypothesis.
10. What is ANOVA?

Basic Inferential Applications **177**

## Chapter Nine Problems

1. Compute a 95% confidence interval using the data provided for this problem. Compare the difference using 90% and 98%.
2. Translate the following statements into a statistical hypothesis study:
   a. The ice cream at this store tastes better than the ice cream from the other store.
   b. This car consumes more fuel than the other car.
   c. This process wastes more than the proposed process.
   d. This material can withstand a greater temperature than the other material.
   e. Using the additive changes the consistency of the fluid.
3. What area in the normal curve corresponds to a z of 0.00, 1.29, 1.65, 2.06, and 2.33?
4. What z corresponds to an area of 0.50, 1.75, 2.50, and 3.00?
5. What would the t value be for a one-tail test with $\alpha = 0.05$ and n= 15?
6. What would the t value be for a one-tail test with $\alpha = 0.01$ and n= 7?
7. What would the t value be for a two-tail test with $\alpha = 0.05$ and n= 15?
8. What would the t value be for a two-tail test with $\alpha = 0.01$ and n= 7?
9. In an ANOVA, the F ratio was calculated as 2.84. With $\alpha = 0.05$ and the degrees of freedom 8 and 20, would the null hypothesis be rejected or retained?

# TEN
## Additional Uses of Statistics in Industry

Thus far, this text has focused on statistics commonly used in quality applications. These methods represent the majority of the ones used in manufacturing and other industrial settings. However, there are other less common applications for statistics in industry. The basics for these applications are the same, including basic probability applications, measures of central tendency, and measures of dispersion. The distributions are also the same, including both discrete and continuous distributions such as binomial, Poisson, normal, t, and F. Although the applications are a little different, the previous chapter will prepare the reader to appreciate them.

This chapter begins with the use of statistics for research and development (R&D) The applications next shift to areas that would normally follow R&D, such as market analysis and forecasting, and then on to operations uses such as inventory and scheduling.

The chapter concludes with other types of industry and institutions, including construction, health, education, and the government. Understandably, there are still other applications of statistics not discussed. Perhaps this chapter can serve as a starting point for applying statistics toward developing a culture of quality and continual improvement in those other industries and organizations.

## Product and System Design

The purpose of R&D is to test a possible new product — typically in the form of a prototype — and refine that product's development into an item that the plant is able to manufacture. Additionally, information from the potential market is collected and integrated into the product throughout the design and development process. At the beginning of the process, the quality of the final product is determined so that it meets the needs of the consumer. Research into reliability and lifetime estimates of the product provide further development information and innovation. For the most part, the basic statistics covered in Chapter Nine are those typically used in R&D. From a broad perspective, these statistics include anything from basic discrete statistics and continuous statistics, to inferential statistics. More complicated statistics are also used in R&D, but go beyond this text.

Several additional tools work independently or in conjunction with the statistics gathered throughout the R&D process. They include Quality Function Deployment and Design for Manufacturing (QFD/DFM), Ishikawa or fishbone diagrams, and reliability estimations.

## QFD/DFM

QFD maps issues that are called the *voice of the customer*; it then compares these issues with the technical requirements to fulfill customer needs. Many qualities of a particular product are correlated with the other qualities of that product. In some cases, improving one quality issue decreases the quality of another element of the same product. For example, if customers want a safe tire, they may have to give up a little of how that tire wears. Some want both. To determine which elements to emphasize, the design team asks customers to quantify the importance of safety and wear by sampling the customers and determining the statistics. If customers ultimately prefer safety, efforts center around designing something safe with the trade-off being wear. This approach for organizing the customer's voice is done in QFD with what is called a *House of Quality*. Figure 10.1 shows a simplified house of quality for the tire example.

In this example, the voice of the customer covers safety, quiet, cost, and wear. This is sometimes referred to as the *what* portion of the house of quality. The ranking of each of these indicates safety the most important to the customer followed by quiet, cost, and wear. The voice of the technician represents technical issues determined to control these customer issues; these technical issues are sometimes referred to as the *how* in the house of quality. Just above the voice of the technician is the roof of the house showing correlation between those technical issues. A positive correlation means as one issue increases, the correlated issue increases. A negative correlation means the opposite: as one issue increases, the correlated issue decreases. No correlation means there is no effect from one to the other. In this case, the only positive correlation was the fact that as the tread thickness increases, so does the cost of materials.

In the house portion of the matrix, 5 indicates high relativity, and 1 indicates small relativity. These relativities are multiplied by the customer ranks, making a score of importance that are shown in the lower portion of the split box in the house. The importance scales are tallied through the matrix for an overall importance factor. This is where the design decision is made to emphasize a certain quality and minimize another. In this case, the design would more closely follow what the customer wants if more emphasis was placed on tread thickness.

To reiterate, this example of a house of quality is a very simple one. They can become quite complicated. Search the internet for more complicated examples to further understand this important tool. Houses of quality can also be used to compare competition. Some matrices include graphs to visualize these comparisons, helping to focus on areas where efforts will improve quality.

QFD begins with the house of quality. Most of the learning and decisions regarding quality are made during this phase. Completing the deployment, however, requires

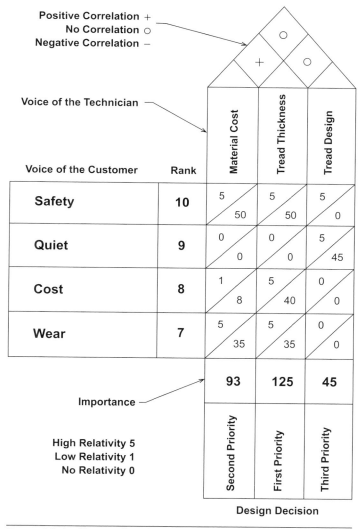

FIGURE 10.1  House of quality regarding tires

looking further at these quality focus areas and how they affect manufacturing. In the end, the manufacturing engineers and technicians must find a way to make the product efficiently with minimal cost while maintaining the need of the customer. Typically this is done through a process called Design for Manufacturing (DFM). Think of the QFD as four phases beginning with the house of quality. The house of quality is where information is processed in planning the design of the product. The second phase is the actual design based on this information. The third phase is using DFM to plan the production process. The final phase is controlling and monitoring the process for quality, possible improvements, and failure prevention. This phase includes the SPC methods discussed in earlier chapters.

## Ishikawa Diagram

The Ishikawa Diagram (named after its developer Kaoru Ishikawa), commonly called a fishbone diagram, is a technique used to visualize cause and effects. It is useful in determining where quality problems originate and can reveal a path of correction. In other words, it can help reveal the unnatural variance within a process causing an inconsistency or quality issue. Typically, in manufacturing, there are six parts (or main bones) in the diagram including machine, method, material, manpower, measurement, and Mother Nature. These are sometimes referred to as the 6Ms; they represent possible sources of a problem, defect, unnatural variance, or some other issue. Through analysis, these issues can be mitigated and quality can improve. Figure 10.2 depicts a typical fishbone diagram aimed at finding the cause of a process defect.

## Life Cycle and Reliability

Reliability is the probability of the product functioning as it is supposed to for a given length of time. This statistic, which can be used to estimate the life cycle of a product, has three major phases. The first phase is the new product that could be defective. Therefore, for a period of time, the probability of failure is elevated and the reliability is low. The second phase represents making it through the initial phase and shows a reduction in the probability of failure, increasing the reliability. Once the product reaches a certain time limit, either by design or by nature, the probability of failure once again increases. Figure 10.3, typically called a *bathtub curve,* shows how

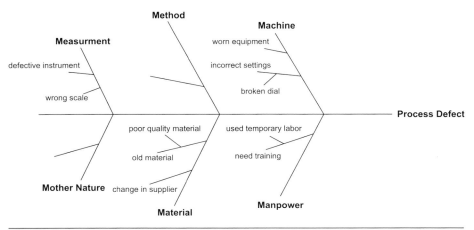

FIGURE 10.2   Typical Ishikawa (Fishbone)

## Additional Uses of Statistics in Industry

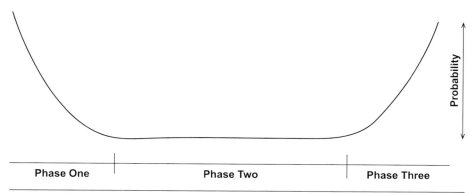

FIGURE 10.3  Typical life cycle curve showing probability of failure through three phases

these probabilities behave throughout the life of the product. Analytically, the probability of failure at any given time (point on the horizontal line of the bathtub chart) is considered:

$$P = e^{-T/MTBF} \qquad (10.1)$$

where e = 2.7183, T represents time before failure, and MTBF represents the mean time before failure.

Proof of this equation goes beyond the scope of this text. However, by collecting an adequate number of historical data you can see how reliability can be expressed and how it can help in the innovation of a product by decreasing the probability of failure over time. Failures in phase one are approximated with negative exponential distributions (equation 10.1). Failures in phase three can be approximated with the normal distribution as follows:

$$Z = T - \bar{x}_{wot}/s_{wot} \qquad (10.2)$$

where

$$wot = \text{wear out time}$$

## DOE

R&D studies have the potential to become quite complicated. The way to handle this is through organization. For many years, statistical experimentation has been standardized with methods called Design of Experiments (DOE). These methods are planned to test relationships of different variance sources — depending on the kind of variance source and the relationship between these sources.

These tests can quickly become too complicated for the level of this text. But a familiarity with their designs will help when you experience them in application.

Let's consider three of the more common DOE methods: the Completely Randomized Design (CR-p), the Randomized Block Design (RB-p), and the Latin Square Design (LS-p).

In Chapter Nine, the experiments were based on a single treatment. The t-test experiment with the dielectric strength of the pc board involved a simple comparison between one group and another. The F-test experiment regarding a concrete plasticizer for increased flow was necessary to avoid multiple t-tests. The plasticizer experiment used one treatment (the plasticizer from four different sources); the variance was analyzed between the groups and within the groups. Each mortar subject was completely randomized (CR) with one treatment (the plasticizer) and subsequently four treatment levels (the plasticizer from the four different vendors). Thus, the experiment had a "CR-4" design. Table 9.4 in Chapter Nine shows the CR-4 design with S1, S2, S3, and S4 being the four treatment levels that produce the subsequent subject results.

Suppose the concrete casting company now wanted to apply the plasticizer to more than one type of concrete use in the plant. They will still use the one treatment with four treatment levels (one plasticizer from each of the four venders). However, the subjects will now be separated into three groups of 8,000 psi concrete, 10,000 psi concrete, and 12,000 psi concrete to ascertain any differences psi might make. The resulting array looks like Table 10.1 and represents a Randomized Block (RB) design with one treatment and four treatment levels (RB-4). The *block* refers to separating the different psi concrete groups from the others.

Not only will this design test the difference between the plasticizers, it will also test if there is a difference between psi blocks. In other words, the hypotheses are:

$H_o: \mu_1 = \mu_2 = \mu_3 = \mu_4$ (for the four plasticizer treatment levels)

$H_o: \mu_1 = \mu_2 = \mu_3$ (for the three blocks regarding psi)

The f-ratio can still be used, but partitioning the sums of squares (SS) is slightly different than what was used in Chapter Nine for finding a significant f-ratio and for the CR-4 above. The TSS is divided into the SSt (for the treatment), the SSb (for the blocks), and the SSr (residual). Two f-ratios are calculated — the first is SSt/SSr to determine significance for the treatment and the second is SSb/SSr regarding significance for the blocks. All other parts of the hypotheses tests remain the same.

To further complicate the experiment, the company realizes they acquire cement from three different suppliers located in different parts of the country to make their concrete. Although the concrete is structurally similar, the plant has noticed differences in the past depending on the supplier. For simplicity, the treatment level (the plasticizers) has been reduced to three rather than four. Now in addition to three treatment levels and the three blocks, the company wants to determine any difference because of three cement suppliers. This analysis can be done using the Latin Square Design or, in this case, an LS-3 design. The tabulated design is represented in the 3 by 3 matrix (sometimes called an *orthogonal array*) in Table 10.2 where the blocks occupy the rows, the cement suppliers occupy columns, and each of the three plasticizer treatment levels are accounted for in each of the matrix cells. Each plasticizer treatment cell contains four data. Note how each treatment appears only once in each row and each column. There

TABLE 10.1  RB-4 Design regarding Concrete Plasticizer Treatment and PSI Blocks

|    | 8000 psi | | | | 10000 psi | | | | 12000 psi | | | |
|----|------|------|------|------|------|------|------|------|------|------|------|------|
|    | 10.9 | 11.0 | 11.4 | 11.8 | 11.3 | 11.0 | 10.2 | 10.8 | 11.0 | 12.7 | 12.6 | 10.0 |
| S1 |      |      |      |      |      |      |      |      |      |      |      |      |
| S2 | 11.0 | 10.0 | 11.4 | 10.6 | 11.3 | 11.4 | 11.8 | 10.8 | 10.3 | 12.9 | 10.5 | 12.5 |
| S3 | 11.6 | 12.2 | 11.6 | 11.0 | 11.9 | 12.0 | 12.2 | 11.0 | 11.8 | 10.0 | 10.7 | 10.6 |
| S4 | 11.3 | 11.1 | 11.4 | 11.3 | 11.4 | 11.7 | 11.0 | 11.6 | 12.7 | 11.2 | 10.1 | 10.0 |

TABLE 10.2  LS-3 Design regarding Concrete Plasticizer Treatment, PSI Blocks, and Concrete Supplier

|    | C1 | C2 | C3 |
|----|----|----|----|
| B1 | S1 | S2 | S3 |
| B2 | S2 | S3 | S1 |
| B3 | S3 | S1 | S2 |

are now three hypotheses to be tested: one on plasticizer treatments, one on psi blocks, and one on the cement suppliers.

The three experimental designs described in this section form the basis for other experimental designs that quickly become more complicated as variables and treatments are added.

## Taguchi

The Taguchi method of experimentation became popular in the 1980s and 1990s. It was developed by Genichi Taguchi in the 1950s because traditional DOE is not clearly suited for real-time production scenarios. The Taguchi method, which focuses on target values in consideration of several variances that can enter the production process, is used to improve both the processes and the product. Typically it is popular in industry because it reduces the number of tests needed to affect improvement in the system, and therefore costs as well. By looking at target values and concentrating on the few variances responsible for attaining those targets, the Taguchi method is sometimes referred to as *Robust* design. In other words, the design will serve to eliminate other variances while testing more apparent variances. These variances are typically referred to as *nuisance variables*. The Taguchi method of experimental design finds the quickest path to optimal operating parameters.

The actual statistics involved in the study and their analyses are the same used in DOE and other statistical studies. The results of the experiment are analyzed and implemented into the process system. Success and failure are measured by the *loss function*. According to this measure, as the variance affecting quality increases, the actual loss or poor quality continue to cost both in money and in consequences for the whole of society, as well as within operations. These results emphasize the importance of controlling variance in the operation process system, even to a minimal level.

Typical general steps to conduct a Taguchi experiment are as follows:

1. Determine the process target.
2. Identify variables and the level of those variables.
3. Create the experiment design (typically an orthogonal array).
4. Conduct the experiment and collect data.
5. Analyze those data to determine what to implement in the process to best attain the process target.

## Market Analysis

In the development of a product, knowing what the customer wants is of prime importance. The importance of customer awareness — or market analysis — is central to most basic philosophies in business and industry. From an operations standpoint, market analysis reveals quality goals, features, price targets, and a multitude of other information used to design both the product and the process. The product is based on *what* the customer wants and the process, to a certain extent, is based on *how much* the customer wants.

Each of these customer-wanted items of information can be analyzed using statistics. One useful statistical tool not yet discussed is forecasting, which is an estimate of what may be in the future given a set of current conditions or variables collected from the past. Forecasting can help industry determine how much product to make, stocking levels, projected sales, and capacity. The discussion will focus on three common forms of forecasting using in industry: the moving average, the weighted moving average and general linear regression. General linear regression can become quite complicated, but in its simplest form can help you understand the principal behind even the complicated designs.

**Moving Average.** The moving average is based on current or very recent past data; it calculates an average for those data using Equation 2.1 from Chapter Two. Suppose a sales team looks at the last three days of sales for a particular item. Three days prior, 249 items were sold; the day before yesterday 254 were sold; and yesterday 260 items were sold. The team can average these figures and forecast what will be sold today — a projection of 254.34 items. There are some problems with this forecast, however. It does not account for the increasing trend. The team can argue (and rightfully so) the data are showing an increased trend; therefore the forecast of 254+ could result in not having enough items to sell for the day.

**Weighted Moving Average.** To alleviate some of the limitations of the moving average, the sales team can give more importance to the most recent information. In other words, older data will have less weight than more recent data. Using Equation 4.1 from Chapter Four, the team determines weights for each day's data. For example, 249 (found three days prior) times 0.2, 254 (from the day before yesterday) times 0.3, and 260 (from yesterday) times 0.5 now calculates to a weighted moving average of 256. This calculation serves as a partial correction with regard to an increased trend, but will not fully correct the problem if the last three data actually represent an upward trend. These methods are hit and miss to some extent; they may be more suited for a known cyclic pattern (in this case, three days). If this pattern is truly cyclic — in other words, the data increased and decreased cyclically every three days — it provides an accurate means of forecasting. There are other ways to estimate this type of pattern, such as exponential smoothing, but the same issue will occur. When a trend is suspected in the data, the forecasting method should switch to some sort of regression to identify that trend. Most regression goes beyond the scope of this text, but the following example describes the most basic type.

# 188   Chapter Ten

**General Linear Regression.** The general linear regression is calculated using what is called *Least Squares*. This regression provides both an analytical approach and a visual approach for understanding the trend or lack of a trend. The basic principal is similar to a house with a roof. The house sits on the ground and the roof has some degree of a slope. This slope shows the trend. Determine the slope, then use it to compute future values.

Analytically, this type of forecasting identifies a change in the dependent variable given a change in the independent variable. For example, as time (the independent variable) increases, sales (the dependent variable) increase as per the slope (how slanted the roof is). The independent variable is located on the horizontal portion of a graph and the dependent variable is located on the vertical portion of the graph. See Figure 10.4 for a typical sales scenario. In this case, the slope (the roof's slant) is 0.6. Therefore, for every increase in the independent variable (period) there is a 0.6 increase in the dependent variable (sales). This pattern shows an upward trend and suggests sales are increasing. For forecasting purposes, this information can be used to determine not only future sale (to an extent), but also an indication that business can continue.

To calculate future sales forecasts, use what is called the *prediction equation* as follows:

$$\hat{y} = \beta_1 + \beta_0 x \qquad (10.3)$$

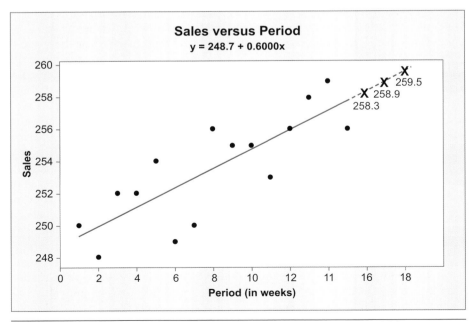

FIGURE 10.4   Sales vs. Period showing forecasts for periods 16, 17, and 18

## Additional Uses of Statistics in Industry 189

where β1 is the intercept and $β_0$ is the slope. The hat symbol over the y simply means that value is predicted and the x is the forecasted period in the future. In the case of Figure 10.4, the prediction equation is

$$\hat{y} = 248.7 + 0.6000x$$

To forecast the next three periods, use equation 10.3 as follows:

$$\hat{y} = 248.7 + 0.6000(16) = 258.3$$
$$\hat{y} = 248.7 + 0.6000(17) = 258.9$$
$$\hat{y} = 248.7 + 0.6000(18) = 259.5$$

These predictions can be seen in Figure 10.4 where the prediction line becomes dotted and the values are represented as shown. The computation of the graph starts with the summary table shown in Table 10.3. Note the period has been transformed into a numerical value for ease of calculation. This forms the x column and becomes the independent variable and the horizontal axis in Figure 10.4. The actual sales data for each of these periods is displayed in column y; the two remaining columns are required for the calculation of the intercept and slope. The intercept is the point on the y-axis (sales) when x is at zero. It represents the beginning point of the slope; therefore, all calculations are added to the intercept to calculate the predicted value.

TABLE 10.3   Sales vs. Period Forecast Summary

| Period | x | y | xy | $x^2$ |
|---|---|---|---|---|
| Wk 1 Mth 1 | 1 | 250 | 250 | 1 |
| Wk 2 Mth 1 | 2 | 248 | 496 | 4 |
| Wk 3 Mth 1 | 3 | 252 | 756 | 9 |
| Wk 4 Mth 1 | 4 | 252 | 1008 | 16 |
| Wk 1 Mth 2 | 5 | 254 | 1270 | 25 |
| Wk 2 Mth 2 | 6 | 249 | 1494 | 36 |
| Wk 3 Mth 2 | 7 | 250 | 1750 | 49 |
| Wk 4 Mth 2 | 8 | 256 | 2048 | 64 |
| Wk 1 Mth 3 | 9 | 255 | 2295 | 81 |
| Wk 2 Mth 3 | 10 | 255 | 2550 | 100 |
| Wk 3 Mth 3 | 11 | 253 | 2783 | 121 |
| Wk 4 Mth 3 | 12 | 256 | 3072 | 144 |
| Wk 1 Mth 4 | 13 | 258 | 3354 | 169 |
| Wk 2 Mth 4 | 14 | 259 | 3626 | 196 |
| Wk 3 Mth 4 | 15 | 256 | 3840 | 225 |
| Σ | 120 | 3803 | 30592 | 1240 |

The calculation of the intercept requires knowledge of the slope, which is calculated as follows:

$$\beta_1 = S_{xy}/S_{xx} \qquad (10.4)$$

where

$$S_{xy} = \Sigma xy - [(\Sigma x)(\Sigma y)/n] \qquad (10.5)$$

and

$$S_{xx} = \Sigma x^2 - [(\Sigma x)^2/n] \qquad (10.6)$$

The intercept is calculated by:

$$\beta_0 = \bar{y} - \beta_1 \bar{x} \qquad (10.7)$$

where $\bar{y}$ and $\bar{x}$ are the averages of the $y$ and $x$ respectively and can be calculated using equation 2.1 from Chapter Two. Continue with the example by referencing information in Table 10.3:

$$S_{xy} = 30592 - (120)(3803)/15 = 176$$

$$S_{xx} = 1240 - (120)^2/15$$

$$= 1240 - 14400/15 = 1240 - 960 = 280$$

so

$$\beta_1 = 176/280 = 0.6$$

and

$$\beta_0 = 253.5 - 0.6(8)$$

$$= 253.5 - 4.8 = 248.7$$

Thus, the prediction equation (equation 10.3) for computing any future value of $x$ in this particular example is:

$$\hat{y} = 248.7 + 0.6000(x)$$

One useful exercise is to calculate the linear regression using these equations on a known outcome. For example, try applying this to 10, 15, 20, 25, and 30. The answer will be 35, 40, and so on. Remember, as with all forecasts and other probability statistics, linear regression provides an estimation of what might happen. A flat slope may indicate business is not growing; however, it does not mean the business will not last. A slope that stays flat for many years would mean many years of gainful employment for those involved. A slope that becomes negative indicates the business may be failing.

Statistics are used in other aspects of design in terms of product, product development, and process design. Location analysis, for example, uses a formula of weighted mean in what is called a *centroid* equation. It can be used for locating entire plants as well as a single machine on the plant floor. Familiarity with these basic statistics will help the reader recognize applications throughout the entire system design.

## System Operation

System operation deals extensively with inventory and scheduling. Statistics are used here as well. In inventory management, safety stock is frequently held to ensure customers get what they want within a certain service level (SL). This SL, which is determined by management, is set in consideration of how critical the item is to the customer and the expense associated with running out of stock. The SL is provided as either a percentage or a coefficient such as 95%. If the SL is 95%, the risk of a stock out is the reciprocal, or 5%. The safety stock is calculated based on past historical data and the SL; it is used in reorder points (ROP), fixed order intervals (FOI), and single period models (SPM).

This use of statistics is applied similarly throughout the entire supply chain. In scheduling and project planning, statistics are used to calculate probabilistic time measurements and sequencing problems. What follows is a brief explanation of how statistics are used in some of the inventory management applications in systems operation.

### ROP

Inventory drops as it is used in the production system. When it reaches the Reorder Point (ROP), staff set in motion the activities necessary for replenishing it, whether the inventory is made in plant or ordered through a supplier. Variation can occur in two areas during this time. First, production can speed up or slow down. In the event it speeds up, there is the possibility (and probability) of a stock out of a required process component. Second, the item can be late in arriving, causing an increase in the lead time (LT) and further concern for production problems. It can also arrive early, causing an increased cost in inventory. Using past data regarding these two variances, namely standard deviation and mean, the probability can be set at the SL for optimal performance.

Suppose a warehouse wishes to find the ROP for a certain brand of chocolate bar with an SL of 90% (meaning a willingness to accept a stock out risk of 10%). The truck that delivers this candy bar takes three days to come once the order is made. It is always on time. Stocking data from the past show the distribution to be normal with a mean of $\bar{x}_d = 50$ and a standard deviation of $\sigma_d = 5$, meaning within a day's time an average of 50 chocolate bars leave the warehouse. The SL is looked up in the area under the normal curve (Appendix B) to acquire the z -score for this scenario. In this case, the z is found to be 1.285. In turn,

$$\text{ROP} = \bar{x}_d \text{LT} + z\, \sigma_d \sqrt{LT} \qquad (10.8)$$

so

$$\text{ROP} = 50(3) + 1.285(5)\sqrt{3}$$
$$= 150 + 11.12 = 161.12$$

This reorder point suggests ordering these chocolate bars when the inventory drops to 162 bars. The answer of 161.12 was rounded to 162 because there is no retail use for 0.12 of a candy bar. This scenario will maintain a 90% in-stock level, meaning 90% of the time the customer will find the candy bar in stock with a 10% risk of being out of stock. Incidentally, the second half of this equation ($z\,\sigma_d\sqrt{LT}$) is the safety stock (11.12 rounded to 12). This is how much stock is required over the nominal amount of 150 bars in order to ensure stock levels in case of increased consumption. Variations of this model include a constant demand with a variable lead time — more likely, both lead time and demand vary. The models change for each case.

## FOI

The Fixed Order Interval (FOI) also relies on safety stock to ensure customer satisfaction to the determined SL. When replenishing inventory using FOI, there is one opportunity per cycle (the order interval) to receive the correct shipment amount. Imagine a pharmacy needing a particular medicine — critical to its customers — that is delivered only once a week. As a result, the SL is set at 98%. With this high SL, the safety stock will be high, resulting in acceptably high inventory costs to minimize any event of running out of this important medicine. The quantity formula within an FOI is given as:

$$Q_{foi} = \bar{x}_d(OI + LT) + z\,\sigma_d\sqrt{OI + LT} - A \qquad (10.9)$$

where, OI is the order interval, and A is the amount on hand at the time of order.

For the current example, the variance associated with the demand is considered normal with a mean of $\bar{x}_d = 42$ tablets per week and a standard deviation of $\sigma_d = 3$. The LT is 3 days and OI is one week. Converting all three factors into equitable time units adjusts quantities as such: LT remains at 3 days, the OI becomes 7 (one week = 7 days), and $\bar{x}_d = 6$ (42 per week/7). Once again, the z-score is found in the area under the normal curve (Appendix B) as 2.55. There are 10 tablets in stock at the time the order is made.

$$Q_{foi} = 42(7+3) + 2.55(3)\sqrt{7+3} - 10$$
$$= 420 + 7.65\sqrt{10} - 10$$
$$= 420 + 24.17 - 10 = 420 + 14.17 = 434.17$$

As in the ROP example, the safety stock is the second half of the equation ($z\,\sigma_d\sqrt{OI + LT}$) and, in this case, is 24.17. This is rounded up to 25. In the end, the pharmacy would order 435 tablets 3 days prior to needing them in stock with 98% confidence customers will have their medicine.

## SPM

The Single Period Model in inventory management provides an optimal stocking level for those items that are considered perishable such as fresh food items, hotel rooms, airline tickets, and newspapers. Overstocking has costs in unsold product that may or may not have a salvage value. Understocking has costs in customer dissatisfaction and perhaps loss of that customer. The SL plays a role in the SPM and is determined by the parameters mentioned above regarding cost rather than by a managerial decision. For the SPM, the SL is calculated using the following:

$$SL = C_s / (C_s + C_e) \qquad (10.10)$$

where $C_s$ is the cost of shortage (unrealized profit) and is calculated as (sale price – cost), and $C_e$ is excess cost and is calculated as (cost – salvage).

For example, an adhesive bonding application in a process requires a cleaning operation with a perishable cleansing solution. The solution is expensive and is mixed every morning. The prepared solution will not last more than 24 hours. Anything left has no salvage and must be discarded. The solution cost $25 per gallon to prepare for the entire working day. After years of data collection, it is determined the distribution is normal, where the operation uses an average of 10 gallons of solution per day with a standard deviation of 3. If the solution runs out, production stops at an average cost of $1500. With

$$C_s = 1500 - 250 = 1250$$
$$C_e = 250 - 0 = 250$$

Therefore,

$$SL = 1250 / (1250+250)$$
$$= 1250/1500 = 0.83$$

This means a stock-out risk of 0.17 is acceptable given the costs involved.

From a discrete perspective, the solution is mixed in gallon containers and used as such. According to the Poisson distribution (Appendix A) with an expected value of 10 (the average per day), cumulative probabilities show the SL of 0.83 fits between 0.791 and 0.864. Following the distribution at those probabilities left to the first column, we see counts of 12 and 13 respectively. This means it would be cheaper to mix 3 more gallons than the anticipated average of 10 per day just to prevent a stock-out. Incidentally, because of the discrete data, the SL gets pushed up to 0.864.

From a continuous perspective, the solution would be mixed in a large drum capable of containing an exact liquid volume. In this case, the stocking level would be calculated as:

$$S_o = \bar{x} + z(\sigma) \tag{10.11}$$

$$S_o = 10 + 0.955(3)$$

$$= 12.865 \text{ gallons, or } 12 \text{ gallons and } 104 \text{ ounces}$$

The z-score was found in Appendix B with an area of 0.83. Note the two solutions give the same results considering only the difference between a discrete distribution and a continuous one.

## Other Applications

There are many other applications of statistics in industry — too many to cover in this text. Statistics in industry have been applied primarily to manufacturing; however, there some potentially beneficial applications elsewhere. Construction is one. With a long tradition of inspection in this industry, efforts should increase to bring manufacturing-style quality control on site. Many of the statistics covered in this text can be applied directly, or with little adaptation, to construction methods. Statistics are already used in project management in estimating probabilities in CPM (Critical Path Method) and PERT (Program Evaluation and Review Technique). But actual control charts, both attribute and continuous, could bring this important industry to a new level of performance.

Historically, statistics were developed and used mostly in insurance and health research. They are appropriately resurfacing in these institutions; but now they are being used for quality control and improvement. Especially in healthcare, considerable efforts are underway to utilize statistical studies to reduce mortality, infections, costs, and to improve service and efficiency. Most of these efforts come in the form of six sigma studies within the realm of lean philosophies.

Another sector that could benefit tremendously from industrial-type statistics is education and other government-controlled institutions. Quality awards (covered more in the next chapter) are already being granted to education institutions in recognition of their improvement in performance and efficiency. Many parts of the government, including the military, could save significant money using these techniques. It's hard to find an entity that is not currently being inundated with lean philosophies. Within these entities, lean requires the use of these statistics for continual improvement of quality and performance.

### Chapter Ten Discussion

1. How are statistics used in product and system design? Name and explain some common tools used for this purpose.
2. What does DOE refer to? What are the three principle types of DOE designs?

3. Explains each: CR-3, RB-4, and LS-3.
4. How does the Taguchi method differ from classic DOE?
5. How are statistics used in a market analysis?
6. Name and explain each: moving average, weighted moving average, and linear regression.
7. How are statistics used in systems operation?

### Chapter Ten Problems

1. Using the Least Squares method of linear regression, calculate the next two periods of capacity based on the following data:
   a. $x$: 1, 2, 3, 4, 5, and 6
   b. $y$: 34, 42, 48, 40, 45, and 52

# ELEVEN

## Quality Awards, Standards, and Quality Management

Although the statistics and other tools introduced in this text are intended primarily to facilitate quality and excellence in performance, they are not the only topic, nor even the first, to assure quality. The statistics used in SPC, acceptance sampling, research, and hypotheses tests fit into a small corner of a much larger room. Foremost is the quality philosophy fostered throughout the entire organization that promotes a common appreciation for the customer and a corporate culture bred into every operation and task providing a quality product or service.

The kind of philosophy suggested here is one that takes years to implement into most industrial settings, especially in those where a change must occur to do so. This is the most common case in the United States. Following WWII, quality was not so much inspected into the product as it was selected into the market. For example, with use of larger than necessary inventories, inspectors separated defective product from acceptable product prior to releasing the product to the consumer. Although this kept defective product from consumers, giving the appearance of a quality product, there was no focus on improving the process to eliminate the initial defective product. In the following years, most manufacturers adopted this as a general practice making quality inspection a function within a *quality control* department.

In *Out of the Crisis* (1982), W. Edwards Deming quoted his colleague Harold Dodge by saying, "you cannot inspect quality into a product." Indeed, the defect was already there by the time the inspection took place. The ideology behind this statement logically suggests that with perfect quality coming from the process, there would be no need for inspection, certainly no need for separation of good and bad, nor data collection, analysis, statistics, or even a quality control department. Alas, the perfect process never has, doesn't, nor ever will exist, as Deming also pointed out. The continual approach to quality pursues as high a quality as possible through planning, studying, analyzing, and controlling. This approach requires a philosophy that centers every activity on the customer and promotes a quality focus from everyone in the plant. It incidentally requires, as Deming also pointed out, a thorough background in statistics and how they are applied in an industrial setting.

Through the second half of the 20th century and continuing on to present day, this philosophy is built on standardization, prizes, methodology, and principal. This chapter provides a brief overview of these ideals that govern, complement, and encourage a drive toward quality and performance excellence. It begins with the major awards such as the Deming Award and the Baldrige Award, and their relationship to other established standards, including ISO and ANSI, and organizations such as ASQ. The chapter then follows with the development and implementation of Lean, JIT, and other cultures conducive to quality at all levels throughout the organization, concluding with a discussion of Six Sigma and the combination of Six Sigma/Lean.

## Awards

### The Deming Prize

At the end of WWII, Deming turned his attention to helping Japanese industry become sustainable in the world market. Primarily, he taught the Japanese principles of quality and work philosophies that fostered a cooperative, customer-driven effort toward continual quality improvement and the perfection of the process through SPC. The methods and content used were nothing new to Deming, but they were new to the Japanese. They were principles developed by Deming and other key quality contemporaries including his strongest influence Walter Shewhart. However, given the Japanese culture of resourcefulness and frugality, these principles made perfect sense when considering ideals such as perfecting a process, reducing waste, increasing efficiency, and reducing costs.

The success of this endeavor speaks for itself. Beginning in 1950, with Deming's assistance, Japanese industry grew to a major power worldwide. Along with this success, came growth and prosperity to the country's economy. Japanese products were traded worldwide and had overcome a powerfully residual reputation of poor quality. By 1980, Japanese products were of the highest quality in the world and remained less expensive than product made in the United States. As a direct result, Japan's economy ultimately rose to the second largest in the world.

In 1951, the Union of Japanese Scientists and Engineers (JUSE) established the Deming Prize in honor of its namesake. This prize, which is the oldest of the quality awards, makes awards in four categories: The Deming Prize for individuals, The Deming Distinguished Service Award for Dissemination and Promotion (Overseas), The Deming Prize (formerly the application prize), and The Deming Grand Prize. Each carries its own criteria, but all are built primarily on what is now called Total Quality Management (TQM), focusing on a continual improvement of quality throughout every process, operation, and task throughout the entire organization. More information regarding this coveted award can be found by visiting the JUSE website.

## The Baldrige Award

With the Japanese making vast improvements in the industrial sector, more countries, including the United States, were importing items made there. This practice caused problems for domestic companies competing against increasing, or perhaps superior quality, and lower prices. In the United States, this issue reached national attention and prompted action from the government.

In the documentary "If Japan Can... Why Can't We?", Deming asserts the United States could match Japan's improvements by using the same principles and philosophies. The *quality revolution* that followed this broadcast ultimately prompted President Reagan and legislators to promote and facilitate quality improvement in the United States and reward extraordinarily successful cases. To do this, Congress created The Malcolm Baldrige National Quality Award (MBNQA), named in honor of the late Secretary of Commerce, Malcolm Baldrige. As Secretary, Baldrige supported a strong sustainable economy through customer-focused quality management.

Each year, up to three awards are issued in each of six categories: Manufacturing, Service, Small Business, Education, Healthcare, and Nonprofit. The seven criteria for performance excellence include successful application of leadership; strategic planning; a market and customer focus; measurement, analysis, and knowledge management; human resource focus; process management; and business/organizational performance results. These awards are similar in to the Deming Prize in terms of Total Quality Management (TQM) and continuous improvement.

## Other Awards

The focus on quality during the 1980s was worldwide. Countries throughout the world created quality awards. Some are given strictly within the country of origin, whereas others consider quality achievements from elsewhere. These awards share a common thread — they promote customer-focused quality improvement and performance excellence. Criteria are basically the same, following a TQM paradigm. Awards that were formed contemporaneously with the MBNQA include the Canada Awards for Excellence Award, the Rajiv Gandhi National Quality Award (India), the EFQM Excellence Award (Europe), the Global Performance Excellence Award (Asia), and the China Quality Award. Currently there are hundreds of recognized quality awards worldwide.

### The Canada Awards for Excellence

The Canada Awards for Excellence (CAE) were formed in 1984 by the Canadian government. Now sponsored by Excellence Canada to promote sustained excellence, they are open to both profit and not-for-profit organizations and governmental entities. There are eleven categories in all covering innovation, wellness, quality, workplace, and customer service.

### Rajiv Gandhi National Quality Award

The Rajiv Gandhi National Quality Award was established in 1991 by the Bureau of Indian Standards (BIS). They recognize industries in India that have improved in quality and performance excellence within several categories, following criteria similar to the other major awards.

### The EFQM Excellence Award
In 1988 the European Foundation for Quality Management was established to provide a model for pursuing quality with the objective of supporting a sustainable European economy. The model was based on the principles of TQM. The EFQM Excellence Award was subsequently given based on how the recipients (both public and private organizations of varying size) fulfilled the expectations of their stakeholders, displayed exceptional performance, and assured this momentum into the future. EFQM also provides training and assessment to help organizations pursue excellence.

### The Global Performance Excellence Award
The Global Performance Excellence Award (GPEA) was founded in 2000 and administered by the Asia Pacific Quality Organization (APQO), then known as the International Asia Pacific Quality Award, with headquarters in the Philippines. It recognizes organizations worldwide for their accomplishments in quality and performance excellence. The award was originally based on the MBNQA format, but now uses the EFQM model — both focused on TQM and performance excellence.

### The China Quality Award
The China Quality Award (CQA) was established in 2001 under the administration of the China Association for Quality (CAQ). The award recognizes both Chinese companies and foreign companies operating in China for performance excellence and principles of TQM. In 2010, the CAQ established a similar award, the China Outstanding Quality Model, to recognize individuals.

## Standards and Societies

The quality societies and associations serve to establish a consortium of professionals who share the same goals, ideology, and stewardship. The terminology may change over the years, but the objective remains the same. These societies strive to improve quality on a continual basis and spread a culture of customer focus and performance excellence. They are tasked with the purpose of developing standards out of these principles and building alliances throughout their realm of operation to maintain this strategy of quality improvement.

These societies or associations may differ from country to country or region, but the message is still the same. They developed early with the terminology of TQM, Continuous Quality Improvement, Kaizen, then later with Performance Excellence. These societies identify new quality trends and innovative ideas leading to higher levels of achievement, then merge these elements into a common standard available for all pertinent organizations. These standards are adopted and finalized by standards organizations such as ANSI in the United States and ISO internationally.

## ISO

The International Organization of Standardization (ISO) began in 1946 in London during an assembly of the Institute of Civil Engineers. Given responsibility for standardizing technology, operations, and management in industry, ISO has since created over 19,000 standards. Today ISO is recognized as the leading international standards organization with 164 member countries and is located in Geneva, Switzerland. Incidentally, the popular name ISO signifies *isos* — Greek for *equal* — providing a connection between regarding the name and the acronym.

The standards most pertinent to quality are the ISO 9000 group. Influenced by several sources including the U.S. Military Standards, the American National Standards Institute (ANSI), and The British Standards Institute (BSI), the ISO 9000 group are the standards most commonly applied worldwide. They have evolved tremendously since then with several versions updating, improving, and adding innovative standards to help industry assure quality in product, system, and service. Currently there are four identifiable standards within the ISO 9000 family: ISO 9001:2008, which covers requirements of a quality management system (QMS); ISO 9000:2005, which deals with concepts and language; ISO 9004:2009, which tries to improve efficiency and effectiveness of the QMS; and ISO 19011:2011, which guides internal and external audits of the QMS.

Although ISO does not provide any certification for any of their standards, ISO 9001:2008 is certifiable through selected independent agencies that, in turn, have been accredited by recognized accreditation boards. The standard requires both internal audits and external audits that document how quality is assured throughout the candidate organization and if that system adequately fulfills expectations of quality and QMS. The new version introduced as ISO 9001:2015 is expected to be released by the end of 2015.

External audits are typically performed by *accredited certification bodies* — they are too numerous to list, but are typically found within the country where the company seeking certification exists. To maintain the standard of certification itself, *accreditation boards* work with these accredited certification bodies to provide guidance in how the certification process should take place. The most prominent accreditation comes from the International Accreditation Forum (IAF). In the United States, IAF members include the American Association for Laboratory Accreditation (A2LA), ANSI–ASQ National Accreditation Board (ANAB), American Nation Standards Institute (ANSI), and the International Accreditation Service (IAS). According to the 2012 ISO survey, more than a million certificates have been issued worldwide regarding ISO 9001 in over 184 countries.

## ANSI

In the United States, the most recognized standards organization is the American National Standards Institute (ANSI). ANSI was formed in 1918 through a consortium of five organizations including the American Institute of Electrical Engineers (now referred to as IEEE), the American Society of Mechanical Engineers (ASME), the American Society of Civil Engineers (ASCE), the American Institute of Mining and Metallurgical Engineers (AIME), and the American Society for Testing Materials (ASTM). Their main objective was to create and maintain standards within the United States and to safeguard these standards promoting international competitiveness and a higher quality of life for U.S. citizens.

ANSI is the established representative of the United States for ISO. They were active in the creation of the ISO and remain active in sharing the interests of ISO. ANSI provides access to ISO standards and works to promote U.S. standards worldwide. ANSI also recognizes certification through third parties similar to ISO.

## ASQ

Unlike ISO and ANSI, the American Society of Quality (ASQ) is an association or, as the name indicates, a society. Founded in 1946, ASQ has approximately 76,000 members in 140 different countries with a vision of making quality a priority nationally and internationally. ASQ also provides quality training, certification, convention, and publication aimed toward promoting quality and quality management. ASQ works closely with ANSI in publishing standards and has formed the ANSI/ASQ National Accreditation Board to facilitate accreditation of the ISO certification boards mentioned above.

## Other Societies

In virtually every country that engages in some sort of manufacturing or industrial activity, there are quality associations or societies. Some of the most notable of there are British Quality Foundation (BQF), European Organization for Quality (EOQ), the German Society for Quality, (DGQ), Japanese Quality Association (JQA), and the China Association of Quality (CAQ). There are too many to list, and each has a unique background; however, the principles and philosophies of TQM and performance excellence are shared among them.

## Six Sigma

Six Sigma was an internal initiative developed at Motorola in the 1980s to increase quality throughout the organization. Aptly named, it makes reference to its statistical background in that many of the resulting projects are aimed toward operations in process and product improvement generally through SPC techniques created by Walter Shewhart in the 1920s. By the late 1980s, significant improvements through use of Six Sigma were obvious and Six Sigma became a thriving culture of quality and performance excellence within Motorola. Bolstered by TQM, continual improvement, and other quality principles, the positive results were indisputable in profit, increase in quality, savings to the customer, and overall efficiency.

The improvements made through the Six Sigma initiative ultimately earned Motorola the Malcolm Baldrige National Quality Award in 1988. Other companies quickly followed including AlliedSignal, Honeywell, and General Electric (GE). GE alone estimates the potential savings between 3 sigma and 6 sigma is 8 to 12 billion USD per year. Since then, Six Sigma has grown tremendously to one of the most common certifications in business and industry. It has emerged as a standard method of improving quality, maximizing profit, and fostering quality friendly cultures throughout the industrial world.

Six Sigma is a project-based improvement technique following a specific pattern within a team environment. It requires support from all levels of management in terms of money as well as sharing in the principles from which Six Sigma was developed. The most successful cases are where the CEO has direct involvement in assuring the team they have support.

The main statistical goal of the Six Sigma team is to reduce the non-conformity rate to 3.4 defects per million opportunities (DPMO). Recall from Chapter Four the area under the curve in the normal distribution and how this area was explained as a probability. The area at $3\sigma$ above the mean was 0.99865 (calculated from negative infinity). This gives a probability of $1 - 0.99856 = 0.00135$, or 1350 ppm. At $6\sigma$, where the area is 0.9999999988, the probability becomes 0.0000000012 or 1.2 ppb (parts-per-billion). If these are manufactured items, this 1.2 ppb is rounded up to 2 ppb. The area under the normal curve that corresponds to 3.4 ppm (either 3 or 4 manufactured items) is actually at $4.5\sigma$ with the area of 0.9999966, or a probability of 0.0000034 (Figure 11.1). Thus, Six Sigma actually holds a goal of $4.5\sigma$ rather than $6\sigma$. This is because over the long term, process means shift to the negative side between $1.4\sigma$

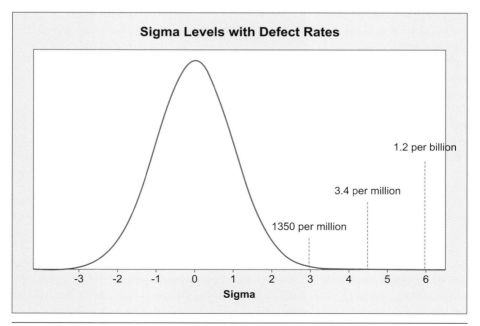

FIGURE 11.1  Sigma levels of 3, 4.5, and 6 (from negative infinity) showing corresponding defect rates

and 1.6σ. During the short term (the Six Sigma improvement phase), process means remain high until assignable variation re-emerges into the process. Motorola called this *long term dynamic mean variation*. Because this 1.5σ process shift is needed to achieve the 3.4 DPMO goal, the process improvement from a short term perspective is 4.5 + 1.5 = 6σ.

The other statistical aspects of Six Sigma are used mostly in the Measure and Analysis phases, described below. Most of the statistics used in Six Sigma are covered in a basic level in this text. Some Six Sigma projects become quite complicated and require more complex statistics. Even determining DPMO can be challenging and require careful consideration to assure the appropriate value. But there is an overall systematic approach to Six Sigma that helps mitigate this complexity. One of two techniques is used depending on the nature of the study. DMADV is used when creating a new system and DMAIC is used when improving an existing system.

## DMAIC

DMAIC is an acronym for Define, Measure, Analyze, Improve, and Control. During the *Define* phase, the problem is clearly stated and understood by all team members. Definition reiterates the voice of the customer by looking at critical quality characteristics, understanding the depths of the problem, determining the variable surrounding the problem, and setting goals for the study. These problems are identified through several sources including management directives, QFD, and benchmarking, but the best problems are considered the *low–hanging fruit*. These are the problems that are easy to see. Once solved, they will reveal other problem areas for further Six Sigma projects.

Once the problem is defined, the team must *Measure* the properties affected by the sources of variance and collect data to analyze. Sometimes this requires measuring the effectiveness of the measurement instrumentation and methods of data collection. The *Gage R&R* is typically done to establish the reliability and repeatability of the measuring system. Once the measurement system is established with assurance of accuracy and precision, the problem is further defined with data that are known to be reliable. During the define phase is where a true picture of the current process becomes evident. To signify the current process, the DPMO, sigma level, $C_{pk}$, or other statistic is typically calculated as sort of a beginning reference point. Later this beginning reference will determine any improvement the study has contributed.

In the *Analyze* phase, the team brainstorms possible root causes of variance and explains that variance with both descriptive and inferential statistical techniques and other analytical tools. They determine how this variance affects the process and reveal problems associated with this variance. Then the team makes plans to remove the root cause of the variance and remedy the problems that are ultimately causing the defects in the process. The tools and statistics at the team's disposal include PERT, CPM, Process Failure Mode and Effects Analysis (PFMEA), Ishikawa Diagrams, 5 Whys (from the Toyota Production System), and a host of visual graphs.

In the *Improve* phase, the root causes of variance found in the analysis phase will be reduced or eliminated. To do this may require any number of changes within the process, including new training, methodology, technology, redesign of the process, a change in material, or a redesign of the product. These implementations are facilitated and monitored by SPC or other statistical experiments to validate the solution.

The final phase, the *Control* phase, is the safeguard against assignable variation re-emerging into the process pushing the DPMO toward $4.5\sigma$. The final phase is the ongoing solution to the original problem; by monitoring and verification the process remains at the specified goal. Typically, in an industrial operations environment, the final phase is done using SPC and other industrial statistics applications. The DPMO, sigma level, or $C_{pk}$ may again be calculated to compare with the original values determined in the measure phase of DMAIC. Ultimately, as the overall goal of Six Sigma is to maximize profit, the financial report brings validation to the management of the organization, a boost in fostering a quality culture throughout the organization, and closure to the project.

## DMADV

DMADV is an acronym for Define, Measure, Analyze, Design, and Verify; it is a popular methodology in Design for Six Sigma (DFSS). It is similar to DMAIC, but supports the development of a new product rather than the improvement of an existing process. The differences are based on customer needs and design. The first three phases are the same as DMAIC; however, the measures are aimed at what the customer wants and the accompanying specifications. The analysis also differs in that it digs into process and design alternatives to meet the customer's wants and needs.

The last two phases differ 1) because a process is in a *design* phase rather than an *improve* phase, and 2) because a new design — whether it is a process, product, or both — requires verification rather than control. Control comes later when implementation efforts shift to DMAIC as opposed to DMADV. The tools and statistics are similar for each of these methods.

## Belts, Certifications, and Organization

The organization of Six Sigma is fairly standardized throughout the industrial world, with each role having specific functions and duties. Most organizations recognize titles associated with Six Sigma to include the Champion, the Master Black Belt (MBB), the Black Belt (BB), and the Green Belt (GB). Some organizations go beyond this list and include Yellow Belts and White Belts.

Champions are involved in every aspect of the Six Sigma project but do not directly participate in the data-driven methodology required throughout the project. They support the team from many different aspects, but most important, they clear the path for the team to operate in the organization without the burden of dealing with territorial disputes, personality conflicts, and hierarchical issues. Champions are empowered by the CEO to facilitate the team within the organization so the team can accomplish their goal. They may or may not even be acquainted with statistics, but they are proponents to team's success. Without the Champions and the full support of the CEO (along with its financial implications), any Six Sigma efforts would be jeopardized.

The remaining members of the team are trained statistically. MBBs are, as the name implies, former BBs with significant experience. They operate as consultants, in most cases giving guidance to BBs and GBs throughout the organization that may involve several projects simultaneously. MBBs do not require formal certification, but advancement to that position does require additional training in more advances statistics and tools required in Six Sigma.

BBs require certification, acquired only after extensive training and documented experience. BBs work full time on the project and lead the team for the length of the project. Their main purpose is to utilize the advice and direction of the MBB and Champion to drive the project through to a successful completion. They direct the efforts of GBs and other subsequent team members.

GBs also require certification. They work part time in the Six Sigma team and are typically busy with the directives of the BBs in collecting data and seeing through various analyses during the project. They will gain valuable experience in this role and eventually (if they so aspire) graduate to a BB position. Certificate programs for both GBs and BBs are numerous and are typically granted by professional organizations and universities. Most of these organizations seek credibility through accreditation.

Six Sigma has gained tremendous attention since its inception; its use is a trend that is increasing rather than waning. Although many proponents emphasize principles of Six Sigma to be data specific (and they are), Six Sigma principles are quite congruent with those mentioned throughout this chapter and text. Those are the principles of TQM, continual improvement, and performance excellence. These project teams strive to improve quality, reduce waste, and increase efficiency. They do so by listening to the voice of the customer. Recently, Six Sigma has been combined innovatively with Lean into Lean Six Sigma, providing a strong level of commitment and contribution to a much broader culture.

## Lean Six Sigma

After World War II, Japan was devastated. The labor force was depleted, and the economy had collapsed. During this time, Eiji Toyoda was forced to make drastic changes to ensure continued business at the Toyota Motor Company. In 1950, Toyoda observed the Ford's River Rouge plant in Dearborn, Michigan where Ford was making several thousand vehicles per day as opposed to Toyota's few hundred per year. Influenced by this visit in several ways, Toyoda implemented many of the ideas he found at Ford with some operational differences geared toward improving efficiency.

Toyota began producing small batches of automobile components and assembling the automobile at a much earlier stage than traditional automobile manufacturing. They *pulled* product through the process as a function of sales rather than *pushing* operations to create inventory that was taking space, time, and unrealized revenue. This approach is now called *Just-in-Time Manufacturing* (JIT); a concept not entirely new to Toyota after Kiichiro Toyoda's visit to Ford in 1929. Toyota separated man's work from machine's work with a concept called *Jidoka* (roughly meaning *automation* with human direction and input). They found machines should be stopped when finished with the operation or when any quality problem arose. This step allowed workers to continue with another machine while the stopped machine was serviced or placed back in operation. Plant personnel used a lighting system called *andon* to show the machine had a problem or was shut down. Within this operational framework, Eiji Toyoda believed that mutual respect among all people in the plant (including between management and workers) would lead to smooth operations and foster a cooperative culture where all employees shared a common pride.

In addition to JIT and Jidoka, Toyota systematically reduced waste and increased efficiency everywhere in the plant. Initially, these improvements occurred mainly in operations; however, they began making improvements in other functions of the organization as well and in management. These improvements became an expected part of the corporate culture throughout all of Toyota where what is called *Kaizen* (meaning "good change") effected continual improvement in reducing waste and increasing efficiency. As a result, Kaizen efforts increased quality, ultimately leading to maximized profit, shorter lead times, and reduced cost to the customer.

Recognizing the importance of these foundations, Eiji Toyoda, Taichi Ohno, and Shigeo Shingo among others at Toyota began documenting the principles and methods into what is known to the rest of the world as Lean Manufacturing. To Toyota, it is simply referred to as The Toyota Production System (TPS). The term *Lean* began in the United States as an observation of Toyota's remarkable achievements. It was initiated by John Krafcik in his 1988 article, "Triumph of the Lean Production System" and was disseminated widely by the 1991 publication of *The Machine That Changed the World: The Story of Lean Production*, by James Womack, Daniel Jones, and Daniel Roos. Lean is known throughout the world as the premier system of reducing cost through reducing waste, increasing quality, and speeding up lead times all within a stable sustainable working environment.

## Lean Principles

The five principles of Lean include identifying value, mapping the value stream, creating flow, establishing pull, and seeking perfection. Identifying value regarding a product or service must begin from the customer's perspective. This is an initial and critical step in lean because every other principle relies on this voice of the customer (VOC). These values established by the customer must be mapped out in the process into what is called a Value Stream Map (VSM). Once the VSM is established and represents the system, all non-value-added tasks must be removed (if possible), creating a continuous link of value-added tasks. The final VSM creates a flow of value throughout the system where the customer no longer pays for non-value-added tasks.

In following a lean approach, and consistent with JIT techniques such as Kanban, you want to establish pull rather than push as a production strategy. Match the output with the demand. Then seek perfection. Note this last principle says seek rather than achieve. Lean principles are not to be seen as unreasonable goals. They are principles to work within and used to build a common culture among all involved in the organization in order to achieve a long-term goal of a high quality stable manufacturing environment.

The short-term goal of Lean is to eliminate waste. The Japanese word for waste is *Muda*. It encompasses all wastes within the working environment. In lean, it centers on those unnecessary entities that add no value to the system, product, or service. Two other sources of waste are described as Muri and Mura. *Muri* (meaning "overburden") is waste from unevenly distributing the work over the entire process. One area is overworked (too little time), producing error, whereas the other area has too much time representing a waste of that resource. *Mura* (meaning "uneven") comes from bottlenecks within the system, a wide product mix, and underutilization of worker resources in flexibility. It also occurs in regard to the unevenness of customer demand. Both Muri and Mura can be controlled by a Lean tool known as Heijunka (meaning "leveling").

There are seven specific sources of Muda. Concentration on each of them can represent countless hours of both Kaizen focus and Lean Six Sigma projects. They include transportation (material handling), unnecessary motion of workers, inventory, waiting, over-processing, over-production, and defects. Note that all of these sources create or represent non-value-added activities or issues.

## Lean Tools

To reduce waste requires tools. A full list of these tools would be too lengthy for this text and would still be incomplete. Several have already been mentioned in this text, such as SPC, and in the last section, such as Heijunka, Andon, Kaizen, Kanban, and Jidoka. Others commonly found in the Lean environment include 5s, SMED, Poka Yoke, PDCA cycle, TPM, and Gemba.

**5S.** The 5S is a way of organizing the plant floor into a safe, easily functioning workplace. They include five words (originally Japanese words) beginning with *s* including sort, set in order, shine, standardize, and sustain.

**SMED.** SMED stands for Single Minute Exchange of Die. It was one of Shigeo Shingo's developments that sped the process of changing dies, allowing small batches and a more even production level over the entire process. Before SMED, changing dies typically took hours to accomplish.

**Poka Yoke.** Poka Yoke means error-proofing. It endeavors to design checks and safeguards into the system to prevent an error from occurring in the first place.

**PDCA.** The PDCA (Plan, Do, Check, and Act) is better known as the Shewhart Cycle. Deming used it extensively in training Japanese manufacturers during the 1950s.

**TPM.** TPM or Total Productive Maintenance is system-wide in maintaining and improving machines and equipment to maximize the function of the process. In turn, it supports and is quite integral to the goal of optimal quality, a rapid and stable process, and ultimate savings to the customer.

**Gemba.** Gemba (or genba) means "the real place". It represents one of the most useful tools in lean by walking to see what is actually happening. Many problems are obvious if you look for them. By simple observation, managers (and line workers too) can find ways to continually improve.

## Six Sigma Combined with Lean

Over the last decade there has been considerable interest generated by combining Lean principles with that of Six Sigma. Lean is a broad corporate philosophy used to foster a culture where a stable system produces product or services of high quality low costs, and rapid lead times. As described in the section above, Lean looks at eliminating waste as a means to achieve this goal. Six Sigma is a prescribed method of quality improvement, as described earlier in this chapter, with the anticipation that once current projects are completed, future projects will be easier to see allowing a continual improvement of the system.

From a fundamental perspective, both Lean and Six Sigma hold the same goals. The difference lies in what these separate philosophies target while achieving this goal. Six Sigma is data intensive and concentrates on reducing variance in the system. Lean is principle specific and concentrates on reducing waste in the system or process. One could argue these targets are the same. After all, waste within the process will introduce assignable variation into the process. Pin-pointing waste, in every sense of form, will present a variance that should be eliminated. Thus, Lean and Six Sigma seem to be a very good match where Lean Six Sigma concentrates its DMAIC methodology on targeting Muda.

### Chapter Eleven Discussion

1. Explain the similarities and differences between the Deming Prize and the MBNQA.
2. What is the underlying common philosophy found in each of the common prizes and awards given for quality and performance excellence?
3. When was and what caused the *quality revolution*?
4. What is ISO and how is it influenced by ASQ and ANSI? When and by whom were these standards created?
5. Describe the certification requirements and procedures for ISO – 9001.
6. Explain the origins of Six Sigma. How is Six Sigma organized and what is the process used to improve quality?
7. What is the difference between DMAIC and DMADV?
8. Explain the foundations of Lean. What are the five principles of Lean?
9. In Lean, explain Muda, Mura, and Muri, and name the seven sources of waste.
10. What are some of the most common tools used by Lean to make continual improvement?
11. What is the link between Six Sigma and Lean?

# Appendix A

# Individual and Cumulative Terms of the Poisson Distribution

## Appendix A

INDIVIDUAL AND CUMULATIVE TERMS of the POISSON DISTRIBUTION (Cumulative on the right, Final Cumulative Rounded)

### λ (μ or np)

| k (c, r, or y) | 0.1 | 0.2 | 0.3 | 0.4 | 0.5 |
|---|---|---|---|---|---|
| 0 | 0.904837 0.904837 | 0.818731 0.818731 | 0.740818 0.740818 | 0.670320 0.670320 | 0.606531 0.606531 |
| 1 | 0.090484 0.995321 | 0.163746 0.982477 | 0.222245 0.963063 | 0.268128 0.938448 | 0.303265 0.909796 |
| 2 | 0.004524 0.999845 | 0.016375 0.998852 | 0.033337 0.996400 | 0.053626 0.992074 | 0.075816 0.985612 |
| 3 | 0.000151 0.999996 | 0.001092 0.999944 | 0.003334 0.999734 | 0.007150 0.999224 | 0.012636 0.998248 |
| 4 | 0.000004 1 | 0.000055 0.999999 | 0.000250 0.999984 | 0.000715 0.999939 | 0.001580 0.999828 |
| 5 |  | 0.000002 1 | 0.000015 0.999999 | 0.000057 0.999996 | 0.000158 0.999986 |
| 6 |  |  | 0.000001 1 | 0.000004 1 | 0.000013 0.999999 |
| 7 |  |  |  |  | 0.000001 1 |

### λ (μ or np)

| k (c, r, or y) | 0.6 | 0.7 | 0.8 | 0.9 | 1.0 |
|---|---|---|---|---|---|
| 0 | 0.548812 0.548812 | 0.496585 0.496585 | 0.449329 0.449329 | 0.406570 0.406570 | 0.367879 0.367879 |
| 1 | 0.329287 0.878099 | 0.347610 0.844195 | 0.359463 0.808792 | 0.365913 0.365913 | 0.367879 0.735758 |
| 2 | 0.098786 0.976885 | 0.121663 0.965858 | 0.143785 0.952577 | 0.164661 0.530574 | 0.183940 0.919698 |
| 3 | 0.019757 0.996642 | 0.028388 0.994246 | 0.038343 0.990920 | 0.049398 0.579972 | 0.061313 0.981011 |
| 4 | 0.002964 0.999606 | 0.004968 0.999214 | 0.007669 0.998589 | 0.011115 0.591087 | 0.015328 0.996339 |
| 5 | 0.000356 0.999962 | 0.000696 0.999910 | 0.001227 0.999816 | 0.002001 0.593088 | 0.003066 0.999405 |
| 6 | 0.000036 0.999998 | 0.000081 0.999991 | 0.000164 0.999980 | 0.000300 0.593388 | 0.000511 0.999916 |
| 7 | 0.000003 1 | 0.000008 0.999999 | 0.000019 0.999999 | 0.000039 0.593427 | 0.000073 0.999989 |
| 8 |  | 0.000001 1 | 0.000002 1 | 0.000004 1 | 0.000009 0.999998 |
| 9 |  |  |  |  | 0.000001 1 |

### λ (μ or np)

| k (c, r, or y) | 1.1 | 1.2 | 1.3 | 1.4 | 1.5 |
|---|---|---|---|---|---|
| 0 | 0.332871 0.332871 | 0.301194 0.301194 | 0.272532 0.272532 | 0.246597 0.246597 | 0.223130 0.223130 |
| 1 | 0.366158 0.699029 | 0.361433 0.662627 | 0.354291 0.626823 | 0.345236 0.591833 | 0.334695 0.557825 |
| 2 | 0.201387 0.900416 | 0.216860 0.879487 | 0.230289 0.857112 | 0.241665 0.833498 | 0.251021 0.808846 |
| 3 | 0.073842 0.974258 | 0.086744 0.966231 | 0.099792 0.956904 | 0.112777 0.946275 | 0.125511 0.934357 |
| 4 | 0.020307 0.994565 | 0.026023 0.992254 | 0.032432 0.989336 | 0.039472 0.985747 | 0.047067 0.981424 |
| 5 | 0.004467 0.999032 | 0.006246 0.998500 | 0.008432 0.997768 | 0.011052 0.996799 | 0.014120 0.995544 |
| 6 | 0.000819 0.999851 | 0.001249 0.999749 | 0.001827 0.999595 | 0.002579 0.999378 | 0.003530 0.999074 |
| 7 | 0.000129 0.999980 | 0.000214 0.999963 | 0.000339 0.999934 | 0.000516 0.999894 | 0.000756 0.999830 |
| 8 | 0.000018 0.999998 | 0.000032 0.999995 | 0.000055 0.999989 | 0.000090 0.999984 | 0.000142 0.999972 |
| 9 | 0.000002 1 | 0.000004 0.999999 | 0.000008 0.999997 | 0.000014 0.999998 | 0.000024 0.999996 |
| 10 |  | 0.000001 1 | 0.000001 1 | 0.000002 1 | 0.000004 1 |

(continued on the next page)

## Individual and Cumulative Terms of the Poisson Distribution

| k | | | $\lambda$ ($\mu$ or np) | | | |
|---|---|---|---|---|---|---|
| | 1.6 | 1.7 | 1.8 | 1.9 | 2.0 | |
| (c, r, or y) | | | | | | |
| 0 | 0.201897 0.201897 | 0.182684 0.182684 | 0.165299 0.165299 | 0.149569 0.149569 | 0.135335 0.135335 | |
| 1 | 0.323034 0.524931 | 0.310562 0.493246 | 0.297538 0.462837 | 0.284180 0.433749 | 0.270671 0.406006 | |
| 2 | 0.258428 0.783359 | 0.263978 0.757224 | 0.267784 0.730621 | 0.269971 0.703720 | 0.270671 0.676677 | |
| 3 | 0.137828 0.921187 | 0.149587 0.906811 | 0.160671 0.891292 | 0.170982 0.874702 | 0.180447 0.857124 | |
| 4 | 0.055131 0.976318 | 0.063575 0.970386 | 0.072302 0.963594 | 0.081216 0.955918 | 0.090224 0.947348 | |
| 5 | 0.017642 0.993960 | 0.021615 0.992001 | 0.026029 0.989623 | 0.030862 0.986780 | 0.036089 0.983437 | |
| 6 | 0.004705 0.998665 | 0.006124 0.998125 | 0.007809 0.997432 | 0.009773 0.996553 | 0.012030 0.995467 | |
| 7 | 0.001075 0.999740 | 0.001487 0.999612 | 0.002008 0.999440 | 0.002653 0.999206 | 0.003437 0.998904 | |
| 8 | 0.000215 0.999955 | 0.000316 0.999928 | 0.000452 0.999892 | 0.000630 0.999836 | 0.000859 0.999763 | |
| 9 | 0.000038 0.999993 | 0.000060 0.999988 | 0.000090 0.999982 | 0.000133 0.999969 | 0.000191 0.999954 | |
| 10 | 0.000006 0.999999 | 0.000010 0.999998 | 0.000016 0.999998 | 0.000025 0.999994 | 0.000038 0.999992 | |
| 11 | 0.000001 1 | 0.000002 1 | 0.000003 1 | 0.000004 0.999999 | 0.000007 0.999999 | |
| 12 | | | | 0.000001 1 | 0.000001 1 | |

| k | | | $\lambda$ ($\mu$ or np) | | | |
|---|---|---|---|---|---|---|
| | 2.1 | 2.2 | 2.3 | 2.4 | 2.5 | |
| (c, r, or y) | | | | | | |
| 0 | 0.122456 0.122456 | 0.110803 0.110803 | 0.100259 0.100259 | 0.090718 0.090718 | 0.082085 0.082085 | |
| 1 | 0.257158 0.379614 | 0.243767 0.354570 | 0.230595 0.330854 | 0.217723 0.308441 | 0.205212 0.287297 | |
| 2 | 0.270016 0.649630 | 0.268144 0.622714 | 0.265185 0.596039 | 0.261268 0.569709 | 0.256516 0.543813 | |
| 3 | 0.189011 0.838641 | 0.196639 0.819353 | 0.203308 0.799347 | 0.209014 0.778723 | 0.213763 0.757576 | |
| 4 | 0.099231 0.937872 | 0.108151 0.927504 | 0.116902 0.916249 | 0.125408 0.904131 | 0.133602 0.891178 | |
| 5 | 0.041677 0.979549 | 0.047587 0.975091 | 0.053775 0.970024 | 0.060196 0.964327 | 0.066801 0.957979 | |
| 6 | 0.014587 0.994136 | 0.017448 0.992539 | 0.020614 0.990638 | 0.024078 0.988405 | 0.027834 0.985813 | |
| 7 | 0.004376 0.998512 | 0.005484 0.998023 | 0.006773 0.997411 | 0.008255 0.996660 | 0.009941 0.995754 | |
| 8 | 0.001149 0.999661 | 0.001508 0.999531 | 0.001947 0.999358 | 0.002477 0.999137 | 0.003106 0.998860 | |
| 9 | 0.000268 0.999929 | 0.000369 0.999900 | 0.000498 0.999856 | 0.000660 0.999797 | 0.000863 0.999723 | |
| 10 | 0.000056 0.999985 | 0.000081 0.999981 | 0.000114 0.999970 | 0.000159 0.999956 | 0.000216 0.999939 | |
| 11 | 0.000011 0.999996 | 0.000016 0.999997 | 0.000024 0.999994 | 0.000035 0.999991 | 0.000049 0.999988 | |
| 12 | 0.000002 1 | 0.000003 0.999999 | 0.000005 0.999999 | 0.000007 0.999998 | 0.000010 0.999998 | |
| 13 | | 0.000001 1 | 0.000001 1 | 0.000001 1 | 0.000002 1 | |

(continued on the next page)

| k | $\lambda$ ($\mu$ or np) | | | | | |
|---|---|---|---|---|---|---|
| (c, r, or y) | 2.6 | 2.7 | 2.8 | 2.9 | 3.0 | |
| 0 | 0.074274 0.074274 | 0.067206 0.067206 | 0.060810 0.060810 | 0.055023 0.055023 | 0.049787 0.049787 | |
| 1 | 0.193111 0.267385 | 0.181455 0.248661 | 0.170268 0.231078 | 0.159567 0.214590 | 0.149361 0.199148 | |
| 2 | 0.251045 0.518430 | 0.244964 0.493625 | 0.238375 0.469453 | 0.231373 0.445963 | 0.224042 0.423190 | |
| 3 | 0.217572 0.736002 | 0.220468 0.714093 | 0.222484 0.691937 | 0.223660 0.669623 | 0.224042 0.647232 | |
| 4 | 0.141422 0.877424 | 0.148816 0.862909 | 0.155739 0.847676 | 0.162154 0.831777 | 0.168031 0.815263 | |
| 5 | 0.073539 0.950963 | 0.080360 0.943269 | 0.087214 0.934890 | 0.094049 0.925826 | 0.100819 0.916082 | |
| 6 | 0.031867 0.982830 | 0.036162 0.979431 | 0.040700 0.975590 | 0.045457 0.971283 | 0.050409 0.966491 | |
| 7 | 0.011836 0.994666 | 0.013948 0.993379 | 0.016280 0.991870 | 0.018832 0.990115 | 0.021604 0.988095 | |
| 8 | 0.003847 0.998513 | 0.004708 0.998087 | 0.005698 0.997568 | 0.006827 0.996942 | 0.008102 0.996197 | |
| 9 | 0.001111 0.999624 | 0.001412 0.999499 | 0.001773 0.999341 | 0.002200 0.999142 | 0.002701 0.998898 | |
| 10 | 0.000289 0.999913 | 0.000381 0.999880 | 0.000496 0.999837 | 0.000638 0.999780 | 0.000810 0.999708 | |
| 11 | 0.000068 0.999981 | 0.000093 0.999974 | 0.000126 0.999963 | 0.000168 0.999948 | 0.000221 0.999929 | |
| 12 | 0.000015 0.999996 | 0.000021 0.999995 | 0.000029 0.999992 | 0.000041 0.999989 | 0.000055 0.999984 | |
| 13 | 0.000003 0.999999 | 0.000004 0.999999 | 0.000006 0.999998 | 0.000009 0.999998 | 0.000013 0.999996 | |
| 14 | 0.000001 1 | 0.000001 1 | 0.000001 1 | 0.000002 1 | 0.000003 0.999999 | |
| 15 | | | | | 0.000001 1 | |

| k | $\lambda$ ($\mu$ or np) | | | | | |
|---|---|---|---|---|---|---|
| (c, r, or y) | 3.1 | 3.2 | 3.3 | 3.4 | 3.5 | |
| 0 | 0.045049 0.045049 | 0.040762 0.040762 | 0.036883 0.036883 | 0.033373 0.033373 | 0.030197 0.030197 | |
| 1 | 0.139653 0.184702 | 0.130439 0.171201 | 0.121714 0.158597 | 0.113469 0.146842 | 0.105691 0.135888 | |
| 2 | 0.216461 0.401163 | 0.208702 0.379903 | 0.200829 0.359426 | 0.192898 0.339740 | 0.184959 0.320847 | |
| 3 | 0.223677 0.624840 | 0.222616 0.602519 | 0.220912 0.580338 | 0.218617 0.558357 | 0.215785 0.536632 | |
| 4 | 0.173350 0.798190 | 0.178093 0.780612 | 0.182252 0.762590 | 0.185825 0.744182 | 0.188812 0.725444 | |
| 5 | 0.107477 0.905667 | 0.113979 0.894591 | 0.120286 0.882876 | 0.126361 0.870543 | 0.132169 0.857613 | |
| 6 | 0.055530 0.961197 | 0.060789 0.955380 | 0.066158 0.949034 | 0.071604 0.942147 | 0.077098 0.934711 | |
| 7 | 0.024592 0.985789 | 0.027789 0.983169 | 0.031189 0.980223 | 0.034779 0.976926 | 0.038549 0.973260 | |
| 8 | 0.009529 0.995318 | 0.011116 0.994285 | 0.012865 0.993088 | 0.014781 0.991707 | 0.016865 0.990125 | |
| 9 | 0.003282 0.998600 | 0.003952 0.998237 | 0.004717 0.997805 | 0.005584 0.997291 | 0.006559 0.996684 | |
| 10 | 0.001018 0.999618 | 0.001265 0.999502 | 0.001557 0.999362 | 0.001899 0.999190 | 0.002296 0.998980 | |
| 11 | 0.000287 0.999905 | 0.000368 0.999870 | 0.000467 0.999829 | 0.000587 0.999777 | 0.000730 0.999710 | |
| 12 | 0.000074 0.999979 | 0.000098 0.999968 | 0.000128 0.999957 | 0.000166 0.999943 | 0.000213 0.999923 | |
| 13 | 0.000018 0.999996 | 0.000024 0.999992 | 0.000033 0.999990 | 0.000043 0.999986 | 0.000057 0.999986 | |
| 14 | 0.000004 0.999999 | 0.000006 0.999998 | 0.000008 0.999998 | 0.000011 0.999994 | 0.000014 0.999994 | |
| 15 | 0.000001 1 | 0.000001 1 | 0.000002 1 | 0.000002 0.999999 | 0.000003 0.999997 | |
| 16 | | | | 0.000001 1 | 0.000002 1 | |

(continued on the next page)

# Individual and Cumulative Terms of the Poisson Distribution

| k | 3.6 | | 3.7 | | 3.8 | | 3.9 | | 4.0 | |
|---|---|---|---|---|---|---|---|---|---|---|
| (c, r, or y) | | | | | $\lambda$ ($\mu$ or np) | | | | | |
| 0 | 0.030197 | 0.030197 | 0.024724 | 0.024724 | 0.022371 | 0.022371 | 0.020242 | 0.020242 | 0.018316 | 0.018316 |
| 1 | 0.105691 | 0.135888 | 0.091477 | 0.116201 | 0.085009 | 0.107380 | 0.078943 | 0.099185 | 0.073263 | 0.091579 |
| 2 | 0.184959 | 0.320847 | 0.169233 | 0.285434 | 0.161517 | 0.268897 | 0.153940 | 0.253125 | 0.146525 | 0.238104 |
| 3 | 0.215785 | 0.536632 | 0.208720 | 0.494154 | 0.204588 | 0.473485 | 0.200122 | 0.453247 | 0.195367 | 0.433471 |
| 4 | 0.188812 | 0.725444 | 0.193066 | 0.687220 | 0.194359 | 0.667844 | 0.195119 | 0.648366 | 0.195367 | 0.628838 |
| 5 | 0.132169 | 0.857613 | 0.142869 | 0.830089 | 0.147713 | 0.815557 | 0.152193 | 0.800559 | 0.156293 | 0.785131 |
| 6 | 0.077098 | 0.934711 | 0.088103 | 0.918192 | 0.093551 | 0.909108 | 0.098925 | 0.899484 | 0.104196 | 0.889327 |
| 7 | 0.038549 | 0.973260 | 0.046568 | 0.964760 | 0.050785 | 0.959893 | 0.055115 | 0.954599 | 0.059540 | 0.948867 |
| 8 | 0.016865 | 0.990125 | 0.021538 | 0.986298 | 0.024123 | 0.984016 | 0.026869 | 0.981468 | 0.029770 | 0.978637 |
| 9 | 0.006559 | 0.996684 | 0.008854 | 0.995152 | 0.010185 | 0.994201 | 0.011643 | 0.993111 | 0.013231 | 0.991868 |
| 10 | 0.002296 | 0.998980 | 0.003276 | 0.998428 | 0.003870 | 0.998071 | 0.004541 | 0.997652 | 0.005292 | 0.997160 |
| 11 | 0.000730 | 0.999710 | 0.001102 | 0.999530 | 0.001337 | 0.999408 | 0.001610 | 0.999262 | 0.001925 | 0.999085 |
| 12 | 0.000213 | 0.999923 | 0.000340 | 0.999870 | 0.000423 | 0.999831 | 0.000523 | 0.999785 | 0.000642 | 0.999727 |
| 13 | 0.000057 | 0.999980 | 0.000097 | 0.999967 | 0.000124 | 0.999955 | 0.000157 | 0.999942 | 0.000197 | 0.999924 |
| 14 | 0.000014 | 0.999994 | 0.000026 | 0.999993 | 0.000034 | 0.999989 | 0.000044 | 0.999986 | 0.000056 | 0.999980 |
| 15 | 0.000003 | 0.999997 | 0.000006 | 0.999999 | 0.000009 | 0.999998 | 0.000011 | 0.999997 | 0.000015 | 0.999995 |
| 16 | 0.000002 | 1 | 0.000001 | 1 | 0.000002 | 1 | 0.000002 | 0.999999 | 0.000004 | 0.999999 |
| 17 | | | | | | | 0.000001 | 1 | 0.000001 | 1 |

| k | 4.1 | | 4.2 | | 4.3 | | 4.4 | | 4.5 | |
|---|---|---|---|---|---|---|---|---|---|---|
| (c, r, or y) | | | | | $\lambda$ ($\mu$ or np) | | | | | |
| 0 | 0.016573 | 0.016573 | 0.014996 | 0.014996 | 0.013569 | 0.013569 | 0.012277 | 0.012277 | 0.011109 | 0.011109 |
| 1 | 0.067948 | 0.084521 | 0.062981 | 0.077977 | 0.058345 | 0.071914 | 0.054020 | 0.066297 | 0.049990 | 0.061099 |
| 2 | 0.139293 | 0.223814 | 0.132261 | 0.210238 | 0.125441 | 0.197355 | 0.118845 | 0.185142 | 0.112479 | 0.173578 |
| 3 | 0.190368 | 0.414182 | 0.185165 | 0.395403 | 0.179799 | 0.377154 | 0.174305 | 0.359447 | 0.168718 | 0.342296 |
| 4 | 0.195127 | 0.609309 | 0.194424 | 0.589827 | 0.193284 | 0.570438 | 0.191736 | 0.551183 | 0.189808 | 0.532104 |
| 5 | 0.160004 | 0.769313 | 0.163316 | 0.753143 | 0.166224 | 0.736662 | 0.168728 | 0.719911 | 0.170827 | 0.702931 |
| 6 | 0.109336 | 0.878649 | 0.114321 | 0.867464 | 0.119127 | 0.855789 | 0.123734 | 0.843645 | 0.128120 | 0.831051 |
| 7 | 0.064040 | 0.942689 | 0.068593 | 0.936057 | 0.073178 | 0.928967 | 0.077775 | 0.921420 | 0.082363 | 0.913414 |
| 8 | 0.032820 | 0.975509 | 0.036011 | 0.972068 | 0.039333 | 0.968300 | 0.042776 | 0.964196 | 0.046329 | 0.959743 |
| 9 | 0.014951 | 0.990460 | 0.016805 | 0.988873 | 0.018793 | 0.987093 | 0.020913 | 0.985109 | 0.023165 | 0.982908 |
| 10 | 0.006130 | 0.996590 | 0.007058 | 0.995931 | 0.008081 | 0.995174 | 0.009202 | 0.994311 | 0.010424 | 0.993332 |
| 11 | 0.002285 | 0.998875 | 0.002695 | 0.998626 | 0.003159 | 0.998333 | 0.003681 | 0.997992 | 0.004264 | 0.997596 |
| 12 | 0.000781 | 0.999656 | 0.000943 | 0.999569 | 0.001132 | 0.999465 | 0.001350 | 0.999342 | 0.001599 | 0.999195 |
| 13 | 0.000246 | 0.999902 | 0.000305 | 0.999874 | 0.000374 | 0.999839 | 0.000457 | 0.999799 | 0.000554 | 0.999749 |
| 14 | 0.000072 | 0.999974 | 0.000091 | 0.999965 | 0.000115 | 0.999954 | 0.000144 | 0.999943 | 0.000178 | 0.999927 |
| 15 | 0.000020 | 0.999994 | 0.000026 | 0.999991 | 0.000033 | 0.999987 | 0.000042 | 0.999985 | 0.000053 | 0.999980 |

(continued on the next page)

| k | | | λ (μ or np) | | |
|---|---|---|---|---|---|
| (c, r, or y) | 4.1 | 4.2 | 4.3 | 4.4 | 4.5 |
| 16 | 0.000005 0.999999 | 0.000007 0.999998 | 0.000009 0.999996 | 0.000012 0.999997 | 0.000015 0.999995 |
| 17 | 0.000001 1 | 0.000002 1 | 0.000002 0.999998 | 0.000002 0.999999 | 0.000004 0.999999 |
| 18 | | | 0.000001 1 | 0.000001 1 | 0.000001 1 |

| k | | | λ (μ or np) | | |
|---|---|---|---|---|---|
| (c, r, or y) | 4.6 | 4.7 | 4.8 | 4.9 | 5.0 |
| 0 | 0.010052 0.010052 | 0.009095 0.009095 | 0.008230 0.008230 | 0.007447 0.007447 | 0.006738 0.006738 |
| 1 | 0.046238 0.056290 | 0.042748 0.051843 | 0.039503 0.047733 | 0.036488 0.043935 | 0.033690 0.040428 |
| 2 | 0.106348 0.162638 | 0.100457 0.152300 | 0.094807 0.142540 | 0.089396 0.133331 | 0.084224 0.124652 |
| 3 | 0.163068 0.325706 | 0.157383 0.309683 | 0.151691 0.294231 | 0.146014 0.279345 | 0.140374 0.265026 |
| 4 | 0.187528 0.513234 | 0.184925 0.494608 | 0.182029 0.476260 | 0.178867 0.458212 | 0.175467 0.440493 |
| 5 | 0.172526 0.685760 | 0.173830 0.668438 | 0.174748 0.651008 | 0.175290 0.633502 | 0.175467 0.615960 |
| 6 | 0.132270 0.818030 | 0.136167 0.804605 | 0.139798 0.790806 | 0.143153 0.776655 | 0.146223 0.762183 |
| 7 | 0.086920 0.904950 | 0.091426 0.896031 | 0.095862 0.886668 | 0.100207 0.876862 | 0.104445 0.866628 |
| 8 | 0.049979 0.954929 | 0.053713 0.949744 | 0.057517 0.944185 | 0.061377 0.938239 | 0.065278 0.931906 |
| 9 | 0.025545 0.980474 | 0.028050 0.977794 | 0.030676 0.974861 | 0.033416 0.971655 | 0.036266 0.968172 |
| 10 | 0.011751 0.992225 | 0.013184 0.990978 | 0.014724 0.989585 | 0.016374 0.988029 | 0.018133 0.986305 |
| 11 | 0.004914 0.997139 | 0.005633 0.996611 | 0.006425 0.996010 | 0.007294 0.995323 | 0.008242 0.994547 |
| 12 | 0.001884 0.999023 | 0.002206 0.998817 | 0.002570 0.998580 | 0.002978 0.998301 | 0.003434 0.997981 |
| 13 | 0.000667 0.999690 | 0.000798 0.999615 | 0.000949 0.999529 | 0.001123 0.999424 | 0.001321 0.999302 |
| 14 | 0.000219 0.999909 | 0.000268 0.999883 | 0.000325 0.999854 | 0.000393 0.999817 | 0.000472 0.999774 |
| 15 | 0.000067 0.999976 | 0.000084 0.999967 | 0.000104 0.999958 | 0.000128 0.999945 | 0.000157 0.999931 |
| 16 | 0.000019 0.999995 | 0.000025 0.999992 | 0.000031 0.999989 | 0.000039 0.999984 | 0.000049 0.999980 |
| 17 | 0.000005 0.999999 | 0.000007 0.999999 | 0.000009 0.999998 | 0.000011 0.999995 | 0.000014 0.999994 |
| 18 | 0.000001 1 | 0.000002 1 | 0.000002 0.999999 | 0.000003 0.999998 | 0.000004 0.999998 |
| 19 | | | 0.000001 1 | 0.000001 1 | 0.000001 1 |

(continued on the next page)

# Individual and Cumulative Terms of the Poisson Distribution

| k | λ (μ or np) | | | | | |
|---|---|---|---|---|---|---|
| (c, r, or y) | 5.5 | 6.0 | 6.5 | 7.0 | 7.5 |
| 0 | 0.004087 0.004087 | 0.002479 0.002479 | 0.001503 0.001503 | 0.000912 0.000912 | 0.000553 0.000553 |
| 1 | 0.022477 0.026564 | 0.014873 0.017352 | 0.009772 0.011275 | 0.006383 0.007295 | 0.004148 0.004701 |
| 2 | 0.061812 0.088376 | 0.044618 0.061970 | 0.031760 0.043035 | 0.022341 0.029636 | 0.015555 0.020256 |
| 3 | 0.113323 0.201699 | 0.089235 0.151205 | 0.068814 0.111849 | 0.052129 0.081765 | 0.038889 0.059145 |
| 4 | 0.155819 0.357518 | 0.133853 0.285058 | 0.111822 0.223671 | 0.091226 0.172991 | 0.072916 0.132061 |
| 5 | 0.171401 0.528919 | 0.160623 0.445681 | 0.145369 0.369040 | 0.127717 0.300708 | 0.109375 0.241436 |
| 6 | 0.157117 0.686036 | 0.160623 0.606304 | 0.157483 0.526523 | 0.149003 0.449711 | 0.136718 0.378154 |
| 7 | 0.123449 0.809485 | 0.137677 0.743981 | 0.146234 0.672757 | 0.149003 0.598714 | 0.146484 0.524638 |
| 8 | 0.084871 0.894356 | 0.103258 0.847239 | 0.118815 0.791572 | 0.130377 0.729091 | 0.137329 0.661967 |
| 9 | 0.051866 0.946222 | 0.068838 0.916077 | 0.085811 0.877383 | 0.101405 0.830496 | 0.114440 0.776407 |
| 10 | 0.028526 0.974748 | 0.041303 0.957380 | 0.055777 0.933160 | 0.070983 0.901479 | 0.085830 0.862237 |
| 11 | 0.014263 0.989011 | 0.022529 0.979909 | 0.032959 0.966119 | 0.045171 0.946650 | 0.058521 0.920758 |
| 12 | 0.006537 0.995548 | 0.011264 0.991173 | 0.017853 0.983972 | 0.026350 0.973000 | 0.036575 0.957333 |
| 13 | 0.002766 0.998314 | 0.005199 0.996372 | 0.008926 0.992898 | 0.014188 0.987188 | 0.021101 0.978434 |
| 14 | 0.001087 0.999401 | 0.002228 0.998600 | 0.004144 0.997042 | 0.007094 0.994282 | 0.011304 0.989738 |
| 15 | 0.000398 0.999799 | 0.000891 0.999491 | 0.001796 0.998838 | 0.003311 0.997593 | 0.005652 0.995390 |
| 16 | 0.000137 0.999936 | 0.000334 0.999825 | 0.000730 0.999568 | 0.001448 0.999041 | 0.002649 0.998039 |
| 17 | 0.000044 0.999980 | 0.000118 0.999943 | 0.000279 0.999847 | 0.000596 0.999637 | 0.001169 0.999208 |
| 18 | 0.000014 0.999994 | 0.000039 0.999982 | 0.000101 0.999948 | 0.000232 0.999869 | 0.000487 0.999695 |
| 19 | 0.000004 0.999998 | 0.000012 0.999994 | 0.000034 0.999982 | 0.000085 0.999954 | 0.000192 0.999887 |
| 20 | 0.000001 1 | 0.000004 0.999998 | 0.000011 0.999993 | 0.000030 0.999984 | 0.000072 0.999959 |
| 21 | | 0.000001 1 | 0.000003 0.999996 | 0.000010 0.999994 | 0.000026 0.999985 |
| 22 | | | 0.000002 0.999998 | 0.000003 0.999997 | 0.000009 0.999994 |
| 23 | | | 0.000001 1 | 0.000002 1 | 0.000003 0.999997 |
| 24 | | | | | 0.000002 1 |

(continued on the next page)

**218** Appendix A

| k (c, r, or y) | 8.0 | | 8.5 | | λ (μ or np) 9.0 | | 9.5 | | 10.0 | |
|---|---|---|---|---|---|---|---|---|---|---|
| 0  | 0.000335 | 0.000335 | 0.000203 | 0.000203 | 0.000123 | 0.000123 | 0.000075 | 0.000075 | 0.000045 | 0.000045 |
| 1  | 0.002684 | 0.003019 | 0.001729 | 0.001932 | 0.001111 | 0.001234 | 0.000711 | 0.000786 | 0.000454 | 0.000499 |
| 2  | 0.010735 | 0.013754 | 0.007350 | 0.009282 | 0.004998 | 0.006232 | 0.003378 | 0.004164 | 0.002270 | 0.002769 |
| 3  | 0.028626 | 0.042380 | 0.020826 | 0.030108 | 0.014994 | 0.021226 | 0.010696 | 0.014860 | 0.007567 | 0.010336 |
| 4  | 0.057252 | 0.099632 | 0.044255 | 0.074363 | 0.033737 | 0.054963 | 0.025403 | 0.040263 | 0.018917 | 0.029253 |
| 5  | 0.091604 | 0.191236 | 0.075233 | 0.149596 | 0.060727 | 0.115690 | 0.048266 | 0.088529 | 0.037833 | 0.067086 |
| 6  | 0.122138 | 0.313374 | 0.106581 | 0.256177 | 0.091090 | 0.206780 | 0.076421 | 0.164950 | 0.063055 | 0.130141 |
| 7  | 0.139587 | 0.452961 | 0.129419 | 0.385596 | 0.117116 | 0.323896 | 0.103714 | 0.268664 | 0.090079 | 0.220220 |
| 8  | 0.139587 | 0.592548 | 0.137508 | 0.523104 | 0.131756 | 0.455652 | 0.123160 | 0.391824 | 0.112599 | 0.332819 |
| 9  | 0.124077 | 0.716625 | 0.129869 | 0.652973 | 0.131756 | 0.587408 | 0.130003 | 0.521827 | 0.125110 | 0.457929 |
| 10 | 0.099262 | 0.815887 | 0.110388 | 0.763361 | 0.118580 | 0.705988 | 0.123502 | 0.645329 | 0.125110 | 0.583039 |
| 11 | 0.072190 | 0.888077 | 0.085300 | 0.848661 | 0.097020 | 0.803008 | 0.106661 | 0.751990 | 0.113736 | 0.696775 |
| 12 | 0.048127 | 0.936204 | 0.060421 | 0.909082 | 0.072765 | 0.875773 | 0.084440 | 0.836430 | 0.094780 | 0.791555 |
| 13 | 0.029616 | 0.965820 | 0.039506 | 0.948588 | 0.050376 | 0.926149 | 0.061706 | 0.898136 | 0.072908 | 0.864463 |
| 14 | 0.016924 | 0.982744 | 0.023986 | 0.972574 | 0.032384 | 0.958533 | 0.041872 | 0.940008 | 0.052077 | 0.916540 |
| 15 | 0.009026 | 0.991770 | 0.013592 | 0.986166 | 0.019431 | 0.977964 | 0.026519 | 0.966527 | 0.034718 | 0.951258 |
| 16 | 0.004513 | 0.996283 | 0.007221 | 0.993387 | 0.010930 | 0.988894 | 0.015746 | 0.982273 | 0.021699 | 0.972957 |
| 17 | 0.002124 | 0.998407 | 0.003610 | 0.996997 | 0.005786 | 0.994680 | 0.008799 | 0.991072 | 0.012764 | 0.985721 |
| 18 | 0.000944 | 0.999351 | 0.001705 | 0.998702 | 0.002893 | 0.997573 | 0.004644 | 0.995716 | 0.007091 | 0.992812 |
| 19 | 0.000397 | 0.999748 | 0.000763 | 0.999465 | 0.001370 | 0.998943 | 0.002322 | 0.998038 | 0.003732 | 0.996544 |
| 20 | 0.000159 | 0.999907 | 0.000324 | 0.999789 | 0.000617 | 0.999560 | 0.001103 | 0.999141 | 0.001866 | 0.998410 |
| 21 | 0.000061 | 0.999968 | 0.000131 | 0.999920 | 0.000264 | 0.999824 | 0.000499 | 0.999640 | 0.000889 | 0.999299 |
| 22 | 0.000022 | 0.999990 | 0.000051 | 0.999971 | 0.000108 | 0.999932 | 0.000215 | 0.999855 | 0.000404 | 0.999703 |
| 23 | 0.000007 | 0.999997 | 0.000019 | 0.999990 | 0.000042 | 0.999974 | 0.000089 | 0.999944 | 0.000176 | 0.999879 |
| 24 | 0.000002 | 0.999999 | 0.000007 | 0.999997 | 0.000016 | 0.999990 | 0.000035 | 0.999979 | 0.000073 | 0.999952 |
| 25 | 0.000001 | 1        | 0.000002 | 0.999999 | 0.000006 | 0.999996 | 0.000013 | 0.999992 | 0.000029 | 0.999981 |
| 26 |          |          | 0.000001 | 1        | 0.000002 | 0.999998 | 0.000005 | 0.999997 | 0.000011 | 0.999992 |
| 27 |          |          |          |          | 0.000001 | 1        | 0.000002 | 0.999999 | 0.000004 | 0.999996 |
| 28 |          |          |          |          |          |          | 0.000001 | 1        | 0.000002 | 0.999998 |
| 29 |          |          |          |          |          |          |          |          | 0.000001 | 1        |

(continued on the next page)

# Individual and Cumulative Terms of the Poisson Distribution

| k (c, r, or y) | 11.0 | | 12.0 | | λ (μ or np) 13.0 | | 14.0 | | 15.0 | |
|---|---|---|---|---|---|---|---|---|---|---|
| 0  | 0.000017 | 0.000017 | 0.000006 | 0.000006 | 0.000002 | 0.000002 | 0.000001 | 0.000001 | 0.000000 | 0.000000 |
| 1  | 0.000184 | 0.000201 | 0.000074 | 0.000080 | 0.000029 | 0.000031 | 0.000012 | 0.000013 | 0.000005 | 0.000005 |
| 2  | 0.001010 | 0.001211 | 0.000442 | 0.000522 | 0.000191 | 0.000222 | 0.000081 | 0.000094 | 0.000034 | 0.000039 |
| 3  | 0.003705 | 0.004916 | 0.001770 | 0.002292 | 0.000828 | 0.001050 | 0.000380 | 0.000474 | 0.000172 | 0.000211 |
| 4  | 0.010189 | 0.015105 | 0.005309 | 0.007601 | 0.002690 | 0.003740 | 0.001331 | 0.001805 | 0.000645 | 0.000856 |
| 5  | 0.022415 | 0.037520 | 0.012741 | 0.020342 | 0.006994 | 0.010734 | 0.003727 | 0.005532 | 0.001936 | 0.002792 |
| 6  | 0.041095 | 0.078615 | 0.025481 | 0.045823 | 0.015153 | 0.025887 | 0.008696 | 0.014228 | 0.004839 | 0.007631 |
| 7  | 0.064577 | 0.143192 | 0.043682 | 0.089505 | 0.028141 | 0.054028 | 0.017392 | 0.031620 | 0.010370 | 0.018001 |
| 8  | 0.088794 | 0.231986 | 0.065523 | 0.155028 | 0.045730 | 0.099758 | 0.030436 | 0.062056 | 0.019444 | 0.037445 |
| 9  | 0.108526 | 0.340512 | 0.087364 | 0.242392 | 0.066054 | 0.165812 | 0.047344 | 0.109400 | 0.032407 | 0.069852 |
| 10 | 0.119378 | 0.459890 | 0.104837 | 0.347229 | 0.085870 | 0.251682 | 0.066282 | 0.175682 | 0.048611 | 0.118463 |
| 11 | 0.119378 | 0.579268 | 0.114368 | 0.461597 | 0.101483 | 0.353165 | 0.084359 | 0.260041 | 0.066287 | 0.184750 |
| 12 | 0.109430 | 0.688698 | 0.114368 | 0.575965 | 0.109940 | 0.463105 | 0.098418 | 0.358459 | 0.082859 | 0.267609 |
| 13 | 0.092595 | 0.781293 | 0.105570 | 0.681535 | 0.109940 | 0.573045 | 0.105989 | 0.464448 | 0.095607 | 0.363216 |
| 14 | 0.072753 | 0.854046 | 0.090489 | 0.772024 | 0.102087 | 0.675132 | 0.105989 | 0.570437 | 0.102436 | 0.465652 |
| 15 | 0.053352 | 0.907398 | 0.072391 | 0.844415 | 0.088475 | 0.763607 | 0.098923 | 0.669360 | 0.102436 | 0.568088 |
| 16 | 0.036680 | 0.944078 | 0.054293 | 0.898708 | 0.071886 | 0.835493 | 0.086558 | 0.755918 | 0.096034 | 0.664122 |
| 17 | 0.023734 | 0.967812 | 0.038325 | 0.937033 | 0.054972 | 0.890465 | 0.071283 | 0.827201 | 0.084736 | 0.748858 |
| 18 | 0.014504 | 0.982316 | 0.025550 | 0.962583 | 0.039702 | 0.930167 | 0.055442 | 0.882643 | 0.070613 | 0.819471 |
| 19 | 0.008397 | 0.990713 | 0.016137 | 0.978720 | 0.027164 | 0.957331 | 0.040852 | 0.923495 | 0.055747 | 0.875218 |
| 20 | 0.004618 | 0.995331 | 0.009682 | 0.988402 | 0.017657 | 0.974988 | 0.028597 | 0.952092 | 0.041810 | 0.917028 |
| 21 | 0.002419 | 0.997750 | 0.005533 | 0.993935 | 0.010930 | 0.985918 | 0.019064 | 0.971156 | 0.029865 | 0.946893 |
| 22 | 0.001210 | 0.998960 | 0.003018 | 0.996953 | 0.006459 | 0.992377 | 0.012132 | 0.983288 | 0.020362 | 0.967255 |
| 23 | 0.000578 | 0.999538 | 0.001574 | 0.998527 | 0.003651 | 0.996028 | 0.007385 | 0.990673 | 0.013280 | 0.980535 |
| 24 | 0.000265 | 0.999803 | 0.000787 | 0.999314 | 0.001977 | 0.998005 | 0.004308 | 0.994981 | 0.008300 | 0.988835 |
| 25 | 0.000117 | 0.999920 | 0.000378 | 0.999692 | 0.001028 | 0.999033 | 0.002412 | 0.997393 | 0.004980 | 0.993815 |
| 26 | 0.000049 | 0.999969 | 0.000174 | 0.999866 | 0.000514 | 0.999547 | 0.001299 | 0.998692 | 0.002873 | 0.996688 |
| 27 | 0.000020 | 0.999989 | 0.000078 | 0.999944 | 0.000248 | 0.999795 | 0.000674 | 0.999366 | 0.001596 | 0.998284 |
| 28 | 0.000008 | 0.999997 | 0.000033 | 0.999977 | 0.000115 | 0.999910 | 0.000337 | 0.999703 | 0.000855 | 0.999139 |
| 29 | 0.000002 | 0.999999 | 0.000014 | 0.999991 | 0.000052 | 0.999962 | 0.000163 | 0.999866 | 0.000442 | 0.999581 |

(continued on the next page)

| k | λ (μ or np) | | | | |
|---|---|---|---|---|---|
| (c, r, or y) | 11.0 | 12.0 | 13.0 | 14.0 | 15.0 |
| 30 | 0.000001 1 | 0.000005 0.999996 | 0.000022 0.999984 | 0.000076 0.999942 | 0.000221 0.999802 |
| 31 | | 0.000002 0.999998 | 0.000009 0.999993 | 0.000034 0.999976 | 0.000107 0.999909 |
| 32 | | 0.000001 1 | 0.000004 0.999997 | 0.000015 0.999991 | 0.000050 0.999959 |
| 33 | | | 0.000001 0.999998 | 0.000006 0.999997 | 0.000023 0.999982 |
| 34 | | | 0.000001 1 | 0.000002 0.999999 | 0.000010 0.999992 |
| 35 | | | | 0.000001 1 | 0.000004 0.999996 |
| 36 | | | | | 0.000002 0.999998 |
| 37 | | | | | 0.000001 0.999999 |

| k | λ (μ or np) | | | | |
|---|---|---|---|---|---|
| (c, r, or y) | 16.0 | 17.0 | 18.0 | 19.0 | 20.0 |
| 0 | 0.0000001 0.0000001 | 0.0000000 0.0000000 | 0.0000000 0.0000000 | 0.0000000 0.0000000 | 0.0000000 0.0000000 |
| 1 | 0.0000018 0.0000019 | 0.0000007 0.0000007 | 0.0000003 0.0000003 | 0.0000001 0.0000001 | 0.0000000 0.0000000 |
| 2 | 0.0000144 0.0000163 | 0.0000060 0.0000067 | 0.0000025 0.0000028 | 0.0000010 0.0000011 | 0.0000004 0.0000004 |
| 3 | 0.0000768 0.0000931 | 0.0000339 0.0000406 | 0.0000148 0.0000176 | 0.0000064 0.0000075 | 0.0000027 0.0000031 |
| 4 | 0.0003073 0.0004004 | 0.0001441 0.0001847 | 0.0000666 0.0000842 | 0.0000304 0.0000379 | 0.0000137 0.0000168 |
| 5 | 0.0009833 0.0013837 | 0.0004898 0.0006745 | 0.0002398 0.0003240 | 0.0001156 0.0001535 | 0.0000550 0.0000718 |
| 6 | 0.0026223 0.0040060 | 0.0013879 0.0020624 | 0.0007195 0.0010435 | 0.0003661 0.0005196 | 0.0001832 0.0002550 |
| 7 | 0.0059937 0.0099997 | 0.0033706 0.0054330 | 0.0018500 0.0028935 | 0.0009937 0.0015133 | 0.0005235 0.0007785 |
| 8 | 0.0119875 0.0219872 | 0.0071625 0.0125955 | 0.0041625 0.0070560 | 0.0023600 0.0038733 | 0.0013087 0.0020872 |
| 9 | 0.0213111 0.0432983 | 0.0135292 0.0261247 | 0.0083251 0.0153811 | 0.0049822 0.0088555 | 0.0029082 0.0049954 |
| 10 | 0.0340977 0.0773960 | 0.0229996 0.0491243 | 0.0149852 0.0303663 | 0.0094662 0.0183217 | 0.0058163 0.0108117 |
| 11 | 0.0495967 0.1269927 | 0.0355448 0.0846691 | 0.0245212 0.0548875 | 0.0163508 0.0346725 | 0.0105751 0.0213868 |
| 12 | 0.0661289 0.1931216 | 0.0503552 0.1350243 | 0.0367818 0.0916693 | 0.0258888 0.0605613 | 0.0176252 0.0390120 |
| 13 | 0.0813894 0.2745110 | 0.0658490 0.2008733 | 0.0509286 0.1425979 | 0.0378374 0.0983987 | 0.0271156 0.0661276 |
| 14 | 0.0930164 0.3675274 | 0.0799596 0.2808329 | 0.0654796 0.2080775 | 0.0513508 0.1497495 | 0.0387366 0.1048642 |
| 15 | 0.0992175 0.4667449 | 0.0906208 0.3714537 | 0.0785755 0.2866530 | 0.0650443 0.2147938 | 0.0516489 0.1565131 |
| 16 | 0.0992175 0.5659624 | 0.0962846 0.4677383 | 0.0883975 0.3750505 | 0.0772401 0.2920339 | 0.0645611 0.2210742 |
| 17 | 0.0933812 0.6593436 | 0.0962846 0.5640229 | 0.0935973 0.4686478 | 0.0863272 0.3783611 | 0.0759542 0.2970284 |
| 18 | 0.0830055 0.7423491 | 0.0909355 0.6549584 | 0.0935973 0.5622451 | 0.0911231 0.4694842 | 0.0843936 0.3814220 |
| 19 | 0.0698994 0.8122485 | 0.0813633 0.7363217 | 0.0886711 0.6509162 | 0.0911231 0.5606073 | 0.0888353 0.4702573 |
| 20 | 0.0559195 0.8681680 | 0.0691588 0.8054805 | 0.0798040 0.7307202 | 0.0865670 0.6471743 | 0.0888353 0.5590926 |
| 21 | 0.0426053 0.9107733 | 0.0559857 0.8614662 | 0.0684035 0.7991237 | 0.0783225 0.7254968 | 0.0846051 0.6436977 |
| 22 | 0.0309857 0.9417590 | 0.0432617 0.9047279 | 0.0559665 0.8550902 | 0.0676422 0.7931390 | 0.0769137 0.7206114 |
| 23 | 0.0215553 0.9633143 | 0.0319760 0.9367039 | 0.0437998 0.8988900 | 0.0558783 0.8490173 | 0.0668815 0.7874929 |
| 24 | 0.0143702 0.9776845 | 0.0226497 0.9593536 | 0.0328499 0.9317399 | 0.0442370 0.8932543 | 0.0557346 0.8432275 |

(continued on the next page)

# Individual and Cumulative Terms of the Poisson Distribution

| k (c, r, or y) | 16.0 | 17.0 | λ (μ or np) 18.0 | 19.0 | 20.0 |
|---|---|---|---|---|---|
| 25 | 0.0091969 0.9868814 | 0.0154018 0.9747554 | 0.0236519 0.9553918 | 0.0336201 0.9268744 | 0.0445876 0.8878151 |
| 26 | 0.0056596 0.9925410 | 0.0100704 0.9848258 | 0.0163744 0.9717662 | 0.0245685 0.9514429 | 0.0342982 0.9221133 |
| 27 | 0.0033539 0.9958949 | 0.0063406 0.9911664 | 0.0109163 0.9826825 | 0.0172890 0.9687319 | 0.0254061 0.9475194 |
| 28 | 0.0019165 0.9978114 | 0.0038497 0.9950161 | 0.0070176 0.9897001 | 0.0117318 0.9804637 | 0.0181472 0.9656666 |
| 29 | 0.0010574 0.9988688 | 0.0022567 0.9972728 | 0.0043558 0.9940559 | 0.0076864 0.9881501 | 0.0125153 0.9781819 |
| 30 | 0.0005639 0.9994327 | 0.0012788 0.9985516 | 0.0026135 0.9966694 | 0.0048680 0.9930181 | 0.0083435 0.9865254 |
| 31 | 0.0002911 0.9997238 | 0.0007013 0.9992529 | 0.0015175 0.9981869 | 0.0029836 0.9960017 | 0.0053829 0.9919083 |
| 32 | 0.0001455 0.9998693 | 0.0003726 0.9996255 | 0.0008536 0.9990405 | 0.0017715 0.9977732 | 0.0033643 0.9952726 |
| 33 | 0.0000706 0.9999399 | 0.0001919 0.9998174 | 0.0004656 0.9995061 | 0.0010200 0.9987932 | 0.0020390 0.9973116 |
| 34 | 0.0000332 0.9999731 | 0.0000960 0.9999134 | 0.0002465 0.9997526 | 0.0005700 0.9993632 | 0.0011994 0.9985110 |
| 35 | 0.0000152 0.9999883 | 0.0000466 0.9999600 | 0.0001268 0.9998794 | 0.0003094 0.9996726 | 0.0006854 0.9991964 |
| 36 | 0.0000067 0.9999950 | 0.0000220 0.9999820 | 0.0000634 0.9999428 | 0.0001633 0.9998359 | 0.0003808 0.9995772 |
| 37 | 0.0000029 0.9999979 | 0.0000101 0.9999921 | 0.0000308 0.9999736 | 0.0000839 0.9999198 | 0.0002058 0.9997830 |
| 38 | 0.0000012 0.9999991 | 0.0000045 0.9999966 | 0.0000146 0.9999882 | 0.0000419 0.9999617 | 0.0001083 0.9998913 |
| 39 | 0.0000005 0.9999996 | 0.0000020 0.9999986 | 0.0000067 0.9999949 | 0.0000204 0.9999821 | 0.0000556 0.9999469 |
| 40 | 0.0000002 0.9999998 | 0.0000008 0.9999994 | 0.0000030 0.9999979 | 0.0000097 0.9999918 | 0.0000278 0.9999747 |
| 41 | 0.0000001 1 | 0.0000003 0.9999997 | 0.0000013 0.9999992 | 0.0000045 0.9999963 | 0.0000135 0.9999882 |
| 42 | | 0.0000001 0.9999998 | 0.0000006 0.9999998 | 0.0000020 0.9999983 | 0.0000065 0.9999947 |
| 43 | | 0.0000001 1 | 0.0000002 0.9999999 | 0.0000009 0.9999992 | 0.0000030 0.9999977 |
| 44 | | | 0.0000001 1 | 0.0000004 0.9999996 | 0.0000014 0.9999991 |
| 45 | | | | 0.0000002 0.9999998 | 0.0000006 0.9999997 |
| 46 | | | | 0.0000001 1 | 0.0000003 0.9999999 |
| 47 | | | | | 0.0000001 1 |

# Appendix B

# Areas Under the Normal Distribution

AREAS UNDER THE NORMAL DISTRIBUTION (the z distribution)

| z | 0.00 | 0.01 | 0.02 | 0.03 | 0.04 | 0.05 | 0.06 | 0.07 | 0.08 | 0.09 |
|---|---|---|---|---|---|---|---|---|---|---|
| -3.4 | 0.00034 | 0.00032 | 0.00031 | 0.00030 | 0.00029 | 0.00028 | 0.00027 | 0.00026 | 0.00025 | 0.00024 |
| -3.3 | 0.00048 | 0.00047 | 0.00045 | 0.00043 | 0.00042 | 0.00040 | 0.00039 | 0.00038 | 0.00036 | 0.00035 |
| -3.2 | 0.00069 | 0.00066 | 0.00064 | 0.00062 | 0.00060 | 0.00058 | 0.00056 | 0.00054 | 0.00052 | 0.00050 |
| -3.1 | 0.00097 | 0.00094 | 0.00090 | 0.00087 | 0.00084 | 0.00082 | 0.00079 | 0.00076 | 0.00074 | 0.00071 |
| -3.0 | 0.00135 | 0.00131 | 0.00126 | 0.00122 | 0.00118 | 0.00114 | 0.00111 | 0.00107 | 0.00104 | 0.00100 |
| -2.9 | 0.00187 | 0.00181 | 0.00175 | 0.00169 | 0.00164 | 0.00159 | 0.00154 | 0.00149 | 0.00144 | 0.00139 |
| -2.8 | 0.00256 | 0.00248 | 0.00240 | 0.00233 | 0.00226 | 0.00219 | 0.00212 | 0.00205 | 0.00199 | 0.00193 |
| -2.7 | 0.00347 | 0.00336 | 0.00326 | 0.00317 | 0.00307 | 0.00298 | 0.00289 | 0.00280 | 0.00272 | 0.00264 |
| -2.6 | 0.00466 | 0.00453 | 0.00440 | 0.00427 | 0.00415 | 0.00402 | 0.00391 | 0.00379 | 0.00368 | 0.00357 |
| -2.5 | 0.00621 | 0.00604 | 0.00587 | 0.00570 | 0.00554 | 0.00539 | 0.00523 | 0.00508 | 0.00494 | 0.00480 |
| -2.4 | 0.00820 | 0.00798 | 0.00776 | 0.00755 | 0.00734 | 0.00714 | 0.00695 | 0.00676 | 0.00657 | 0.00639 |
| -2.3 | 0.01072 | 0.01044 | 0.01017 | 0.00990 | 0.00964 | 0.00939 | 0.00914 | 0.00889 | 0.00866 | 0.00842 |
| -2.2 | 0.01390 | 0.01355 | 0.01321 | 0.01287 | 0.01255 | 0.01222 | 0.01191 | 0.01160 | 0.01130 | 0.01101 |
| -2.1 | 0.01786 | 0.01743 | 0.01700 | 0.01659 | 0.01618 | 0.01578 | 0.01539 | 0.01500 | 0.01463 | 0.01426 |
| -2.0 | 0.02275 | 0.02222 | 0.02169 | 0.02118 | 0.02068 | 0.02018 | 0.01970 | 0.01923 | 0.01876 | 0.01831 |
| -1.9 | 0.02872 | 0.02807 | 0.02743 | 0.02680 | 0.02619 | 0.02559 | 0.02500 | 0.02442 | 0.02385 | 0.02330 |
| -1.8 | 0.03593 | 0.03515 | 0.03438 | 0.03362 | 0.03288 | 0.03216 | 0.03144 | 0.03074 | 0.03005 | 0.02938 |
| -1.7 | 0.04457 | 0.04363 | 0.04272 | 0.04182 | 0.04093 | 0.04006 | 0.03920 | 0.03836 | 0.03754 | 0.03673 |
| -1.6 | 0.05480 | 0.05370 | 0.05262 | 0.05155 | 0.05050 | 0.04947 | 0.04846 | 0.04746 | 0.04648 | 0.04551 |
| -1.5 | 0.06681 | 0.06552 | 0.06426 | 0.06301 | 0.06178 | 0.06057 | 0.05938 | 0.05821 | 0.05705 | 0.05592 |
| -1.4 | 0.08076 | 0.07927 | 0.07780 | 0.07636 | 0.07493 | 0.07353 | 0.07215 | 0.07078 | 0.06944 | 0.06811 |
| -1.3 | 0.09680 | 0.09510 | 0.09342 | 0.09176 | 0.09012 | 0.08851 | 0.08691 | 0.08534 | 0.08379 | 0.08226 |
| -1.2 | 0.11507 | 0.11314 | 0.11123 | 0.10935 | 0.10749 | 0.10565 | 0.10383 | 0.10204 | 0.10027 | 0.09853 |
| -1.1 | 0.13567 | 0.13350 | 0.13136 | 0.12924 | 0.12714 | 0.12507 | 0.12302 | 0.12100 | 0.11900 | 0.11702 |

(continued on the next page)

# Areas Under the Normal Distribution

| z | 0.00 | 0.01 | 0.02 | 0.03 | 0.04 | 0.05 | 0.06 | 0.07 | 0.08 | 0.09 |
|---|---|---|---|---|---|---|---|---|---|---|
| -1.0 | 0.15866 | 0.15625 | 0.15386 | 0.15151 | 0.14917 | 0.14686 | 0.14457 | 0.14231 | 0.14007 | 0.13786 |
| -0.9 | 0.18406 | 0.18141 | 0.17879 | 0.17619 | 0.17361 | 0.17106 | 0.16853 | 0.16602 | 0.16354 | 0.16109 |
| -0.8 | 0.21186 | 0.20897 | 0.20611 | 0.20327 | 0.20045 | 0.19766 | 0.19489 | 0.19215 | 0.18943 | 0.18673 |
| -0.7 | 0.24196 | 0.23885 | 0.23576 | 0.23270 | 0.22965 | 0.22663 | 0.22363 | 0.22065 | 0.21770 | 0.21476 |
| -0.6 | 0.27425 | 0.27093 | 0.26763 | 0.26435 | 0.26109 | 0.25785 | 0.25463 | 0.25143 | 0.24825 | 0.24510 |
| -0.5 | 0.30854 | 0.30503 | 0.30153 | 0.29806 | 0.29460 | 0.29116 | 0.28774 | 0.28434 | 0.28096 | 0.27760 |
| -0.4 | 0.34458 | 0.34090 | 0.33724 | 0.33360 | 0.32997 | 0.32636 | 0.32276 | 0.31918 | 0.31561 | 0.31207 |
| -0.3 | 0.38209 | 0.37828 | 0.37448 | 0.37070 | 0.36693 | 0.36317 | 0.35942 | 0.35569 | 0.35197 | 0.34827 |
| -0.2 | 0.42074 | 0.41683 | 0.41294 | 0.40905 | 0.40517 | 0.40129 | 0.39743 | 0.39358 | 0.38974 | 0.38591 |
| -0.1 | 0.46017 | 0.45620 | 0.45224 | 0.44828 | 0.44433 | 0.44038 | 0.43644 | 0.43251 | 0.42858 | 0.42465 |
| -0.0 | 0.50000 | 0.49601 | 0.49202 | 0.48803 | 0.48405 | 0.48006 | 0.47608 | 0.47210 | 0.46812 | 0.46414 |
| 0.0 | 0.50000 | 0.50399 | 0.50798 | 0.51197 | 0.51595 | 0.51994 | 0.52392 | 0.52790 | 0.53188 | 0.53586 |
| 0.1 | 0.53983 | 0.54380 | 0.54776 | 0.55172 | 0.55567 | 0.55962 | 0.56356 | 0.56749 | 0.57142 | 0.57535 |
| 0.2 | 0.57926 | 0.58317 | 0.58706 | 0.59095 | 0.59483 | 0.59871 | 0.60257 | 0.60642 | 0.61026 | 0.61409 |
| 0.3 | 0.61791 | 0.62172 | 0.62552 | 0.62930 | 0.63307 | 0.63683 | 0.64058 | 0.64431 | 0.64803 | 0.65173 |
| 0.4 | 0.65542 | 0.65910 | 0.66276 | 0.66640 | 0.67003 | 0.67364 | 0.67724 | 0.68082 | 0.68439 | 0.68793 |
| 0.5 | 0.69146 | 0.69497 | 0.69847 | 0.70194 | 0.70540 | 0.70884 | 0.71226 | 0.71566 | 0.71904 | 0.72240 |
| 0.6 | 0.72575 | 0.72907 | 0.73237 | 0.73565 | 0.73891 | 0.74215 | 0.74537 | 0.74857 | 0.75175 | 0.75490 |
| 0.7 | 0.75804 | 0.76115 | 0.76424 | 0.76730 | 0.77035 | 0.77337 | 0.77637 | 0.77935 | 0.78230 | 0.78524 |
| 0.8 | 0.78814 | 0.79103 | 0.79389 | 0.79673 | 0.79955 | 0.80234 | 0.80511 | 0.80785 | 0.81057 | 0.81327 |
| 0.9 | 0.81594 | 0.81859 | 0.82121 | 0.82381 | 0.82639 | 0.82894 | 0.83147 | 0.83398 | 0.83646 | 0.83891 |
| 1.0 | 0.84134 | 0.84375 | 0.50798 | 0.84849 | 0.85083 | 0.85314 | 0.85543 | 0.85769 | 0.85993 | 0.86214 |

(continued on the next page)

| z | 0.00 | 0.01 | 0.02 | 0.03 | 0.04 | 0.05 | 0.06 | 0.07 | 0.08 | 0.09 |
|---|------|------|------|------|------|------|------|------|------|------|
| 1.1 | 0.86433 | 0.86650 | 0.86864 | 0.87076 | 0.87286 | 0.87493 | 0.87698 | 0.87900 | 0.88100 | 0.88298 |
| 1.2 | 0.88493 | 0.88686 | 0.88877 | 0.89065 | 0.89251 | 0.89435 | 0.89617 | 0.89796 | 0.89973 | 0.90147 |
| 1.3 | 0.90320 | 0.90490 | 0.90658 | 0.90824 | 0.90988 | 0.91149 | 0.91309 | 0.91466 | 0.91621 | 0.91774 |
| 1.4 | 0.91924 | 0.92073 | 0.92220 | 0.92364 | 0.92507 | 0.92647 | 0.92785 | 0.92922 | 0.93056 | 0.93189 |
| 1.5 | 0.93319 | 0.93448 | 0.93574 | 0.93699 | 0.93822 | 0.93943 | 0.94062 | 0.94179 | 0.94295 | 0.94408 |
| 1.6 | 0.94520 | 0.94630 | 0.94738 | 0.94845 | 0.94950 | 0.95053 | 0.95154 | 0.95254 | 0.95352 | 0.95449 |
| 1.7 | 0.95543 | 0.95637 | 0.95728 | 0.95818 | 0.95907 | 0.95994 | 0.96080 | 0.96164 | 0.96246 | 0.96327 |
| 1.8 | 0.96407 | 0.96485 | 0.96562 | 0.96638 | 0.96712 | 0.96784 | 0.96856 | 0.96926 | 0.96995 | 0.97062 |
| 1.9 | 0.97128 | 0.97193 | 0.97257 | 0.97320 | 0.97381 | 0.97441 | 0.97500 | 0.97558 | 0.97615 | 0.97670 |
| 2.0 | 0.97725 | 0.97778 | 0.50798 | 0.97882 | 0.97932 | 0.97982 | 0.98030 | 0.98077 | 0.98124 | 0.98169 |
| 2.1 | 0.98214 | 0.98257 | 0.98300 | 0.98341 | 0.98382 | 0.98422 | 0.98461 | 0.98500 | 0.98537 | 0.98574 |
| 2.2 | 0.98610 | 0.98645 | 0.98679 | 0.98713 | 0.98745 | 0.98778 | 0.98809 | 0.98840 | 0.98870 | 0.98899 |
| 2.3 | 0.98928 | 0.98956 | 0.98983 | 0.99010 | 0.99036 | 0.99061 | 0.99086 | 0.99111 | 0.99134 | 0.99158 |
| 2.4 | 0.99180 | 0.99202 | 0.99224 | 0.99245 | 0.99266 | 0.99286 | 0.99305 | 0.99324 | 0.99343 | 0.99361 |
| 2.5 | 0.99379 | 0.99396 | 0.99413 | 0.99430 | 0.99446 | 0.99461 | 0.99477 | 0.99492 | 0.99506 | 0.99520 |
| 2.6 | 0.99534 | 0.99547 | 0.99560 | 0.99573 | 0.99585 | 0.99598 | 0.99609 | 0.99621 | 0.99632 | 0.99643 |
| 2.7 | 0.99653 | 0.99664 | 0.99674 | 0.99683 | 0.99693 | 0.99702 | 0.99711 | 0.99720 | 0.99728 | 0.99736 |
| 2.8 | 0.99744 | 0.99752 | 0.99760 | 0.99767 | 0.99774 | 0.99781 | 0.99788 | 0.99795 | 0.99801 | 0.99807 |
| 2.9 | 0.99813 | 0.99819 | 0.99825 | 0.99831 | 0.99836 | 0.99841 | 0.99846 | 0.99851 | 0.99856 | 0.99861 |
| 3.0 | 0.99865 | 0.99869 | 0.50798 | 0.99878 | 0.99882 | 0.99886 | 0.99889 | 0.99893 | 0.99896 | 0.99900 |
| 3.1 | 0.99903 | 0.99906 | 0.99910 | 0.99913 | 0.99916 | 0.99918 | 0.99921 | 0.99924 | 0.99926 | 0.99929 |
| 3.2 | 0.99931 | 0.99934 | 0.99936 | 0.99938 | 0.99940 | 0.99942 | 0.99944 | 0.99946 | 0.99948 | 0.99950 |
| 3.3 | 0.99952 | 0.99953 | 0.99955 | 0.99957 | 0.99958 | 0.99960 | 0.99961 | 0.99962 | 0.99964 | 0.99965 |
| 3.4 | 0.99966 | 0.99968 | 0.99969 | 0.99970 | 0.99971 | 0.99972 | 0.99973 | 0.99974 | 0.99975 | 0.99976 |

# Appendix C

# Critical Values of the t-Distribution

CRITICAL VALUES of the t-DISTRIBUTION

| df | α  0.1 | 0.05 | 0.025 | 0.01 | 0.005 | 0.001 |
|---|---|---|---|---|---|---|
| 1 | 3.078 | 6.314 | 12.706 | 31.821 | 63.657 | 318.309 |
| 2 | 1.886 | 2.920 | 4.303 | 6.965 | 9.925 | 22.327 |
| 3 | 1.638 | 2.353 | 3.182 | 4.541 | 5.841 | 10.215 |
| 4 | 1.533 | 2.132 | 2.776 | 3.747 | 4.604 | 7.173 |
| 5 | 1.476 | 2.015 | 2.571 | 3.365 | 4.032 | 5.893 |
| 6 | 1.440 | 1.943 | 2.447 | 3.143 | 3.707 | 5.208 |
| 7 | 1.415 | 1.895 | 2.365 | 2.998 | 3.499 | 4.785 |
| 8 | 1.397 | 1.860 | 2.306 | 2.896 | 3.355 | 4.501 |
| 9 | 1.383 | 1.833 | 2.262 | 2.821 | 3.250 | 4.297 |
| 10 | 1.372 | 1.812 | 2.228 | 2.764 | 3.169 | 4.144 |
| 11 | 1.363 | 1.796 | 2.201 | 2.718 | 3.106 | 4.025 |
| 12 | 1.356 | 1.782 | 2.179 | 2.681 | 3.055 | 3.930 |
| 13 | 1.350 | 1.771 | 2.160 | 2.650 | 3.012 | 3.852 |
| 14 | 1.345 | 1.761 | 2.145 | 2.624 | 2.977 | 3.787 |
| 15 | 1.341 | 1.753 | 2.131 | 2.602 | 2.947 | 3.733 |
| 16 | 1.337 | 1.746 | 2.120 | 2.583 | 2.921 | 3.686 |
| 17 | 1.333 | 1.740 | 2.110 | 2.567 | 2.898 | 3.646 |
| 18 | 1.330 | 1.734 | 2.101 | 2.552 | 2.878 | 3.610 |
| 19 | 1.328 | 1.729 | 2.093 | 2.539 | 2.861 | 3.579 |
| 20 | 1.325 | 1.725 | 2.086 | 2.528 | 2.845 | 3.552 |
| 21 | 1.323 | 1.721 | 2.080 | 2.518 | 2.831 | 3.527 |
| 22 | 1.321 | 1.717 | 2.074 | 2.508 | 2.819 | 3.505 |
| 23 | 1.319 | 1.714 | 2.069 | 2.500 | 2.807 | 3.485 |
| 24 | 1.318 | 1.711 | 2.064 | 2.492 | 2.797 | 3.467 |
| 25 | 1.316 | 1.708 | 2.060 | 2.485 | 2.787 | 3.450 |

(continued on the next page)

# Critical Values of the t-Distribution

| df | α   0.1 | 0.05 | 0.025 | 0.01 | 0.005 | 0.001 |
|-----|------|-------|-------|-------|-------|-------|
| 26  | 1.315 | 1.706 | 2.056 | 2.479 | 2.779 | 3.435 |
| 27  | 1.314 | 1.703 | 2.052 | 2.473 | 2.771 | 3.421 |
| 28  | 1.313 | 1.701 | 2.048 | 2.467 | 2.763 | 3.408 |
| 29  | 1.311 | 1.699 | 2.045 | 2.462 | 2.756 | 3.396 |
| 30  | 1.310 | 1.697 | 2.042 | 2.457 | 2.750 | 3.385 |
| 31  | 1.309 | 1.696 | 2.040 | 2.453 | 2.744 | 3.375 |
| 32  | 1.309 | 1.694 | 2.037 | 2.449 | 2.738 | 3.365 |
| 33  | 1.308 | 1.692 | 2.035 | 2.445 | 2.733 | 3.356 |
| 34  | 1.307 | 1.691 | 2.032 | 2.441 | 2.728 | 3.348 |
| 35  | 1.306 | 1.690 | 2.030 | 2.438 | 2.724 | 3.340 |
| 40  | 1.303 | 1.684 | 2.021 | 2.423 | 2.704 | 3.307 |
| 50  | 1.299 | 1.676 | 2.009 | 2.403 | 2.678 | 3.261 |
| 60  | 1.296 | 1.671 | 2.000 | 2.390 | 2.660 | 3.232 |
| 120 | 1.289 | 1.658 | 1.980 | 2.358 | 2.617 | 3.160 |

# Appendix D

# Critical Values of the F-Distribution

# Critical Values of the F-Distribution

CRITICAL VALUES of the F-DISTRIBUTION for $\alpha = 0.10$

| $df_w$ \ $df_b$ | 1 | 2 | 3 | 4 | 5 | 6 | 7 | 8 | 9 | 10 | 15 | 20 |
|---|---|---|---|---|---|---|---|---|---|---|---|---|
| 1 | 39.863 | 49.500 | 53.593 | 55.833 | 57.240 | 58.204 | 58.906 | 59.439 | 59.858 | 60.195 | 61.220 | 61.740 |
| 2 | 8.526 | 9.000 | 9.162 | 9.243 | 9.293 | 9.326 | 9.349 | 9.367 | 9.381 | 9.392 | 9.425 | 9.441 |
| 3 | 5.538 | 5.462 | 5.391 | 5.343 | 5.309 | 5.285 | 5.266 | 5.252 | 5.240 | 5.230 | 5.200 | 5.184 |
| 4 | 4.545 | 4.325 | 4.191 | 4.107 | 4.051 | 4.010 | 3.979 | 3.955 | 3.936 | 3.920 | 3.870 | 3.844 |
| 5 | 4.060 | 3.780 | 3.619 | 3.520 | 3.453 | 3.405 | 3.368 | 3.339 | 3.316 | 3.297 | 3.238 | 3.207 |
| 6 | 3.776 | 3.463 | 3.289 | 3.181 | 3.108 | 3.055 | 3.014 | 2.983 | 2.958 | 2.937 | 2.871 | 2.836 |
| 7 | 3.589 | 3.257 | 3.074 | 2.961 | 2.883 | 2.827 | 2.785 | 2.752 | 2.725 | 2.703 | 2.632 | 2.595 |
| 8 | 3.458 | 3.113 | 2.924 | 2.806 | 2.726 | 2.668 | 2.624 | 2.589 | 2.561 | 2.538 | 2.464 | 2.425 |
| 9 | 3.360 | 3.006 | 2.813 | 2.693 | 2.611 | 2.551 | 2.505 | 2.469 | 2.440 | 2.416 | 2.340 | 2.298 |
| 10 | 3.285 | 2.924 | 2.728 | 2.605 | 2.522 | 2.461 | 2.414 | 2.377 | 2.347 | 2.323 | 2.244 | 2.201 |
| 11 | 3.225 | 2.860 | 2.660 | 2.536 | 2.451 | 2.389 | 2.342 | 2.304 | 2.274 | 2.248 | 2.167 | 2.123 |
| 12 | 3.177 | 2.807 | 2.606 | 2.480 | 2.394 | 2.331 | 2.283 | 2.245 | 2.214 | 2.188 | 2.105 | 2.060 |
| 13 | 3.136 | 2.763 | 2.560 | 2.434 | 2.347 | 2.283 | 2.234 | 2.195 | 2.164 | 2.138 | 2.053 | 2.007 |
| 14 | 3.102 | 2.726 | 2.522 | 2.395 | 2.307 | 2.243 | 2.193 | 2.154 | 2.122 | 2.095 | 2.010 | 1.962 |
| 15 | 3.073 | 2.695 | 2.490 | 2.361 | 2.273 | 2.208 | 2.158 | 2.119 | 2.086 | 2.059 | 1.972 | 1.924 |
| 16 | 3.048 | 2.668 | 2.462 | 2.333 | 2.244 | 2.178 | 2.128 | 2.088 | 2.055 | 2.028 | 1.940 | 1.891 |
| 17 | 3.026 | 2.645 | 2.437 | 2.308 | 2.218 | 2.152 | 2.102 | 2.061 | 2.028 | 2.001 | 1.912 | 1.862 |
| 18 | 3.007 | 2.624 | 2.416 | 2.286 | 2.196 | 2.130 | 2.079 | 2.038 | 2.005 | 1.977 | 1.887 | 1.837 |
| 19 | 2.990 | 2.606 | 2.397 | 2.266 | 2.176 | 2.109 | 2.058 | 2.017 | 1.984 | 1.956 | 1.865 | 1.814 |
| 20 | 2.975 | 2.589 | 2.380 | 2.249 | 2.158 | 2.091 | 2.040 | 1.999 | 1.965 | 1.937 | 1.845 | 1.794 |
| 21 | 2.961 | 2.575 | 2.365 | 2.233 | 2.142 | 2.075 | 2.023 | 1.982 | 1.948 | 1.920 | 1.827 | 1.776 |
| 22 | 2.949 | 2.561 | 2.351 | 2.219 | 2.128 | 2.060 | 2.008 | 1.967 | 1.933 | 1.904 | 1.811 | 1.759 |
| 23 | 2.937 | 2.549 | 2.339 | 2.207 | 2.115 | 2.047 | 1.995 | 1.953 | 1.919 | 1.890 | 1.796 | 1.744 |
| 24 | 2.927 | 2.538 | 2.327 | 2.195 | 2.103 | 2.035 | 1.983 | 1.941 | 1.906 | 1.877 | 1.783 | 1.730 |
| 25 | 2.918 | 2.528 | 2.317 | 2.184 | 2.092 | 2.024 | 1.971 | 1.929 | 1.895 | 1.866 | 1.771 | 1.718 |
| 26 | 2.909 | 2.519 | 2.307 | 2.174 | 2.082 | 2.014 | 1.961 | 1.919 | 1.884 | 1.855 | 1.760 | 1.706 |
| 27 | 2.901 | 2.511 | 2.299 | 2.165 | 2.073 | 2.005 | 1.952 | 1.909 | 1.874 | 1.845 | 1.749 | 1.695 |
| 28 | 2.894 | 2.503 | 2.291 | 2.157 | 2.064 | 1.996 | 1.943 | 1.900 | 1.865 | 1.836 | 1.740 | 1.685 |
| 29 | 2.887 | 2.495 | 2.283 | 2.149 | 2.057 | 1.988 | 1.935 | 1.892 | 1.857 | 1.827 | 1.731 | 1.676 |
| 30 | 2.881 | 2.489 | 2.276 | 2.142 | 2.049 | 1.980 | 1.927 | 1.884 | 1.849 | 1.819 | 1.722 | 1.667 |
| 40 | 2.835 | 2.440 | 2.226 | 2.091 | 1.997 | 1.927 | 1.873 | 1.829 | 1.793 | 1.763 | 1.662 | 1.605 |
| 60 | 2.791 | 2.393 | 2.177 | 2.041 | 1.946 | 1.875 | 1.819 | 1.775 | 1.738 | 1.707 | 1.603 | 1.543 |
| 120 | 2.748 | 2.347 | 2.130 | 1.992 | 1.896 | 1.824 | 1.767 | 1.722 | 1.684 | 1.652 | 1.545 | 1.482 |

(continued on the next page)

CRITICAL VALUES of the F-DISTRIBUTION for $\alpha = 0.05$

| $df_w$ | $df_b$ 1 | 2 | 3 | 4 | 5 | 6 | 7 | 8 | 9 | 10 | 15 | 20 |
|---|---|---|---|---|---|---|---|---|---|---|---|---|
| 1 | 161.448 | 199.500 | 215.707 | 224.583 | 230.162 | 233.986 | 236.768 | 238.883 | 240.543 | 241.882 | 245.950 | 248.013 |
| 2 | 18.513 | 19.000 | 19.164 | 19.247 | 19.296 | 19.330 | 19.353 | 19.371 | 19.385 | 19.396 | 19.429 | 19.446 |
| 3 | 10.128 | 9.552 | 9.277 | 9.117 | 9.013 | 8.941 | 8.887 | 8.845 | 8.812 | 8.786 | 8.703 | 8.660 |
| 4 | 7.709 | 6.944 | 6.591 | 6.388 | 6.256 | 6.163 | 6.094 | 6.041 | 5.999 | 5.964 | 5.858 | 5.803 |
| 5 | 6.608 | 5.786 | 5.409 | 5.192 | 5.050 | 4.950 | 4.876 | 4.818 | 4.772 | 4.735 | 4.619 | 4.558 |
| 6 | 5.987 | 5.143 | 4.757 | 4.534 | 4.387 | 4.284 | 4.207 | 4.147 | 4.099 | 4.060 | 3.938 | 3.874 |
| 7 | 5.591 | 4.737 | 4.347 | 4.120 | 3.972 | 3.866 | 3.787 | 3.726 | 3.677 | 3.637 | 3.511 | 3.445 |
| 8 | 5.318 | 4.459 | 4.066 | 3.838 | 3.687 | 3.581 | 3.500 | 3.438 | 3.388 | 3.347 | 3.218 | 3.150 |
| 9 | 5.117 | 4.256 | 3.863 | 3.633 | 3.482 | 3.374 | 3.293 | 3.230 | 3.179 | 3.137 | 3.006 | 2.936 |
| 10 | 4.965 | 4.103 | 3.708 | 3.478 | 3.326 | 3.217 | 3.135 | 3.072 | 3.020 | 2.978 | 2.845 | 2.774 |
| 11 | 4.844 | 3.982 | 3.587 | 3.357 | 3.204 | 3.095 | 3.012 | 2.948 | 2.896 | 2.854 | 2.719 | 2.646 |
| 12 | 4.747 | 3.885 | 3.490 | 3.259 | 3.106 | 2.996 | 2.913 | 2.849 | 2.796 | 2.753 | 2.617 | 2.544 |
| 13 | 4.667 | 3.806 | 3.411 | 3.179 | 3.025 | 2.915 | 2.832 | 2.767 | 2.714 | 2.671 | 2.533 | 2.459 |
| 14 | 4.600 | 3.739 | 3.344 | 3.112 | 2.958 | 2.848 | 2.764 | 2.699 | 2.646 | 2.602 | 2.463 | 2.388 |
| 15 | 4.543 | 3.682 | 3.287 | 3.056 | 2.901 | 2.790 | 2.707 | 2.641 | 2.588 | 2.544 | 2.403 | 2.328 |
| 16 | 4.494 | 3.634 | 3.239 | 3.007 | 2.852 | 2.741 | 2.657 | 2.591 | 2.538 | 2.494 | 2.352 | 2.276 |
| 17 | 4.451 | 3.592 | 3.197 | 2.965 | 2.810 | 2.699 | 2.614 | 2.548 | 2.494 | 2.450 | 2.308 | 2.230 |
| 18 | 4.414 | 3.555 | 3.160 | 2.928 | 2.773 | 2.661 | 2.577 | 2.510 | 2.456 | 2.412 | 2.269 | 2.191 |
| 19 | 4.381 | 3.522 | 3.127 | 2.895 | 2.740 | 2.628 | 2.544 | 2.477 | 2.423 | 2.378 | 2.234 | 2.155 |
| 20 | 4.351 | 3.493 | 3.098 | 2.866 | 2.711 | 2.599 | 2.514 | 2.447 | 2.393 | 2.348 | 2.203 | 2.124 |
| 21 | 4.325 | 3.467 | 3.072 | 2.840 | 2.685 | 2.573 | 2.488 | 2.420 | 2.366 | 2.321 | 2.176 | 2.096 |
| 22 | 4.301 | 3.443 | 3.049 | 2.817 | 2.661 | 2.549 | 2.464 | 2.397 | 2.342 | 2.297 | 2.151 | 2.071 |
| 23 | 4.279 | 3.422 | 3.028 | 2.796 | 2.640 | 2.528 | 2.442 | 2.375 | 2.320 | 2.275 | 2.128 | 2.048 |
| 24 | 4.260 | 3.403 | 3.009 | 2.776 | 2.621 | 2.508 | 2.423 | 2.355 | 2.300 | 2.255 | 2.108 | 2.027 |
| 25 | 4.242 | 3.385 | 2.991 | 2.759 | 2.603 | 2.490 | 2.405 | 2.337 | 2.282 | 2.236 | 2.089 | 2.007 |
| 26 | 4.225 | 3.369 | 2.975 | 2.743 | 2.587 | 2.474 | 2.388 | 2.321 | 2.265 | 2.220 | 2.072 | 1.990 |
| 27 | 4.210 | 3.354 | 2.960 | 2.728 | 2.572 | 2.459 | 2.373 | 2.305 | 2.250 | 2.204 | 2.056 | 1.974 |
| 28 | 4.196 | 3.340 | 2.947 | 2.714 | 2.558 | 2.445 | 2.359 | 2.291 | 2.236 | 2.190 | 2.041 | 1.959 |
| 29 | 4.183 | 3.328 | 2.934 | 2.701 | 2.545 | 2.432 | 2.346 | 2.278 | 2.223 | 2.177 | 2.027 | 1.945 |
| 30 | 4.171 | 3.316 | 2.922 | 2.690 | 2.534 | 2.421 | 2.334 | 2.266 | 2.211 | 2.165 | 2.015 | 1.932 |
| 40 | 4.085 | 3.232 | 2.839 | 2.606 | 2.449 | 2.336 | 2.249 | 2.180 | 2.124 | 2.077 | 1.924 | 1.839 |
| 60 | 4.001 | 3.150 | 2.758 | 2.525 | 2.368 | 2.254 | 2.167 | 2.097 | 2.040 | 1.993 | 1.836 | 1.748 |
| 120 | 3.920 | 3.072 | 2.680 | 2.447 | 2.290 | 2.175 | 2.087 | 2.016 | 1.959 | 1.910 | 1.750 | 1.659 |

(continued on the next page)

# Critical Values of the F-Distribution

CRITICAL VALUES of the F-DISTRIBUTION for $\alpha = 0.01$

| $df_w$ | $df_b$ 1 | 2 | 3 | 4 | 5 | 6 | 7 | 8 | 9 | 10 | 15 | 20 |
|---|---|---|---|---|---|---|---|---|---|---|---|---|
| 1 | 4052.181 | 4999.500 | 5403.352 | 5624.583 | 5763.650 | 5858.986 | 5928.356 | 5981.070 | 6022.473 | 6055.847 | 6157.285 | 6208.730 |
| 2 | 98.503 | 99.000 | 99.166 | 99.249 | 99.299 | 99.333 | 99.356 | 99.374 | 99.388 | 99.399 | 99.433 | 99.449 |
| 3 | 34.116 | 30.817 | 29.457 | 28.710 | 28.237 | 27.911 | 27.672 | 27.489 | 27.345 | 27.229 | 26.872 | 26.690 |
| 4 | 21.198 | 18.000 | 16.694 | 15.977 | 15.522 | 15.207 | 14.976 | 14.799 | 14.659 | 14.546 | 14.198 | 14.020 |
| 5 | 16.258 | 13.274 | 12.060 | 11.392 | 10.967 | 10.672 | 10.456 | 10.289 | 10.158 | 10.051 | 9.722 | 9.553 |
| 6 | 13.745 | 10.925 | 9.780 | 9.148 | 8.746 | 8.466 | 8.260 | 8.102 | 7.976 | 7.874 | 7.559 | 7.396 |
| 7 | 12.246 | 9.547 | 8.451 | 7.847 | 7.460 | 7.191 | 6.993 | 6.840 | 6.719 | 6.620 | 6.314 | 6.155 |
| 8 | 11.259 | 8.649 | 7.591 | 7.006 | 6.632 | 6.371 | 6.178 | 6.029 | 5.911 | 5.814 | 5.515 | 5.359 |
| 9 | 10.561 | 8.022 | 6.992 | 6.422 | 6.057 | 5.802 | 5.613 | 5.467 | 5.351 | 5.257 | 4.962 | 4.808 |
| 10 | 10.044 | 7.559 | 6.552 | 5.994 | 5.636 | 5.386 | 5.200 | 5.057 | 4.942 | 4.849 | 4.558 | 4.405 |
| 11 | 9.646 | 7.206 | 6.217 | 5.668 | 5.316 | 5.069 | 4.886 | 4.744 | 4.632 | 4.539 | 4.251 | 4.099 |
| 12 | 9.330 | 6.927 | 5.953 | 5.412 | 5.064 | 4.821 | 4.640 | 4.499 | 4.388 | 4.296 | 4.010 | 3.858 |
| 13 | 9.074 | 6.701 | 5.739 | 5.205 | 4.862 | 4.620 | 4.441 | 4.302 | 4.191 | 4.100 | 3.815 | 3.665 |
| 14 | 8.862 | 6.515 | 5.564 | 5.035 | 4.695 | 4.456 | 4.278 | 4.140 | 4.030 | 3.939 | 3.656 | 3.505 |
| 15 | 8.683 | 6.359 | 5.417 | 4.893 | 4.556 | 4.318 | 4.142 | 4.004 | 3.895 | 3.805 | 3.522 | 3.372 |
| 16 | 8.531 | 6.226 | 5.292 | 4.773 | 4.437 | 4.202 | 4.026 | 3.890 | 3.780 | 3.691 | 3.409 | 3.259 |
| 17 | 8.400 | 6.112 | 5.185 | 4.669 | 4.336 | 4.102 | 3.927 | 3.791 | 3.682 | 3.593 | 3.312 | 3.162 |
| 18 | 8.285 | 6.013 | 5.092 | 4.579 | 4.248 | 4.015 | 3.841 | 3.705 | 3.597 | 3.508 | 3.227 | 3.077 |
| 19 | 8.185 | 5.926 | 5.010 | 4.500 | 4.171 | 3.939 | 3.765 | 3.631 | 3.523 | 3.434 | 3.153 | 3.003 |
| 20 | 8.096 | 5.849 | 4.938 | 4.431 | 4.103 | 3.871 | 3.699 | 3.564 | 3.457 | 3.368 | 3.088 | 2.938 |
| 21 | 8.017 | 5.780 | 4.874 | 4.369 | 4.042 | 3.812 | 3.640 | 3.506 | 3.398 | 3.310 | 3.030 | 2.880 |
| 22 | 7.945 | 5.719 | 4.817 | 4.313 | 3.988 | 3.758 | 3.587 | 3.453 | 3.346 | 3.258 | 2.978 | 2.827 |
| 23 | 7.881 | 5.664 | 4.765 | 4.264 | 3.939 | 3.710 | 3.539 | 3.406 | 3.299 | 3.211 | 2.931 | 2.781 |
| 24 | 7.823 | 5.614 | 4.718 | 4.218 | 3.895 | 3.667 | 3.496 | 3.363 | 3.256 | 3.168 | 2.889 | 2.738 |
| 25 | 7.770 | 5.568 | 4.675 | 4.177 | 3.855 | 3.627 | 3.457 | 3.324 | 3.217 | 3.129 | 2.850 | 2.699 |
| 26 | 7.721 | 5.526 | 4.637 | 4.140 | 3.818 | 3.591 | 3.421 | 3.288 | 3.182 | 3.094 | 2.815 | 2.664 |
| 27 | 7.677 | 5.488 | 4.601 | 4.106 | 3.785 | 3.558 | 3.388 | 3.256 | 3.149 | 3.062 | 2.783 | 2.632 |
| 28 | 7.636 | 5.453 | 4.568 | 4.074 | 3.754 | 3.528 | 3.358 | 3.226 | 3.120 | 3.032 | 2.753 | 2.602 |
| 29 | 7.598 | 5.420 | 4.538 | 4.045 | 3.725 | 3.499 | 3.330 | 3.198 | 3.092 | 3.005 | 2.726 | 2.574 |
| 30 | 7.562 | 5.390 | 4.510 | 4.018 | 3.699 | 3.473 | 3.304 | 3.173 | 3.067 | 2.979 | 2.700 | 2.549 |
| 40 | 7.314 | 5.179 | 4.313 | 3.828 | 3.514 | 3.291 | 3.124 | 2.993 | 2.888 | 2.801 | 2.522 | 2.369 |
| 60 | 7.077 | 4.977 | 4.126 | 3.649 | 3.339 | 3.119 | 2.953 | 2.823 | 2.718 | 2.632 | 2.352 | 2.198 |
| 120 | 6.851 | 4.787 | 3.949 | 3.480 | 3.174 | 2.956 | 2.792 | 2.663 | 2.559 | 2.472 | 2.192 | 2.035 |

# Appendix E

# Data Sets for *Statistics for Quality Control*

**Data Sets for *Statistics for Quality Control***

Several of the running examples and end-of-chapter problems use data sets that are available online for your use. These examples and problems are listed below. Each set provides data formatted for both Excel® and Minitab®. You can download these data sets at the following URL:

https://new.industrialpress.com/statistics-for-quality-control.html

*Chapter Two*
    PPM Data Set
    End of Chapter Two Problem 5 Data Set

*Chapter Four*
    PPM Data Set
    RC Data Set
    End of Chapter Four Problem 1 Data Set

*Chapter Five*
    Whistles Data Set
    End of Chapter Five Problem 2 Data Set

*Chapter Six*
    End of Chapter Six Problem 1 Data Set

*Chapter Seven*
    Shaft Diameters Data Set
    End of Chapter Seven Problem 1 Data Set

*Chapter Nine*
    Yeast Data Set
    Dielectric Strength Data Set
    Cement Flow Data Set

# INDEX

5 Whys  205
5S  208
80-20 rule  36

abnormally shaped distributions  82–84
acceptance  52, 137–155
Acceptance Quality Level, *see AQL*
Accreditation  201
Accuracy  2–3, 22
Aluminum  13
American National Standards Institute, *see ANSI*
American Society of Quality  202
Amperes  8
analysis of variance  *see ANOVA*
analyze  204
ANOVA  157, 169, 172–176
ANSI  200–202
AOQ  152–153
apothecary pound  7
appropriate scales  14
AQL  140–141, 147, 149–151
area  7
array, ascending  23–24
ASQ  105, 202
assignable variation  87, 124
ATI  152–154
Atom  7
attribute charts  45
attribute type data  5, 87
attributes  15, 57, 87–101
automated inspection  154–155
availability  8
average incoming quality  152–153
average total inspection  152–154
avoirdupois pound  7
awards  198–200
axis  34

Baldrige Award  199, 202
base numbers  4

bathtub curve  182–183
bell shape  28
bias  19, 21, 138-139
bin  24
binomial distributions  54, 56–58, 149, 151
boundary  24, 33
box and whisker plots  33–35
breakdowns  37
brightness  8
British standard scale  6–7
British thermal units (Btus)  12–13

Calibration  22
Calipers  22
Calories  12
Canada Awards  199
Candela  8
Capability  128–133
Capacity  7
cardinal numbers  4
c-charts  91, 98–100, 119
cells  24, 26–29
Celsius scale  11–12
centerline  110, 123
central tendency  18, 24, 63–70, 87
centroid equation  190–191
certification  201
chain sampling  147
chance  47
chemical measures  9–10
chemical resistance  22
China Quality Award  200
Chromatography  14
CI  158–162
circuit  8
class  24
classical probability  43–44
collecting data  18
color  9
combinations  51–52
common cause variation  87

Completely Randomized Design (CR-p)  184
confidence  18, 157
confidence intervals  158–162
conforming  47–49, 57, 91
consistency  22
constants  109
consumers  147–148
continuity  8
continuous distribution  53, 63–84
continuous measures  103–117
continuous quality improvement  200
continuous variable  14
control  205
control charts  45, 87, 89–101, 109, 119–135
control limits  94–97
Coulomb  8
counting  51–53
counting numbers  4
CPM  194, 205
critical path method  194
critical value  163
cubic measures  7
cubit  5
cumulative frequency chart  28
cumulative histogram  28, 31
cumulative probability  61
cups  7
current  8
curves  38–40
customer  180
CV  163
cycles  36, 38
cylinder head  103–116

data
    collection  21–23, 91
    grouped  24–40
    organizing  17–18, 23–40
    raw  23
    tabulation  23–24
data sets  43
decimal scales  6–7
defects  5, 149–150, 203
define  204
degree of freedom  160–161, 168
delivery  8

Deming Prize  198
Deming, Edward  87, 197–198
Demographics  19, 138
Density  20
dependent axis  34
dependent events  49–50
descending array  23–24
descriptive statistics  63, 157
design  179–183
Design for Manufacturing  180–182
Design of Experiments  183–186
deviation  15
DFM  180–182
dial caliper  22
diameter  2, 133–135
dielectric strength  2, 169–172
discrete distributions  43–61
discrete measures  103
discrete variables  14
dispersion  24, 32, 43, 63, 70–75, 87, 113
distribution  28, 32–33, 53, 63–84
distribution curve  38
DMADV  205
DMAIC  204–205
DOE  183–186
double sampling  144–147
dry volume  7

EFQM Excellence Award  200
electrical energy  12
electrical weight  8
elements  9, 22
Empirical Rule  76–78, 158
energy  11–12, 137
enthalpy  12
equipment  137
error  18, 123–124, 165–167
estimations  43
ethics  21
evaluation  46
event  44, 47, 51–53
experimental probability  45
experimental tests  157, 162–176, 183–186
experts  14
exposure  22

F ratio  157, 169, 172, 175, 184
Fahrenheit scale  5, 11–12

# Index

false negative  166
false positive  166
Farad  8
fishbone diagram  182
fixed order intervals (FOI)  191–193
flatness  104
foot  6
forecasting  179, 188
fractional measurements  6
frequency  9, 65–66
frequency charts  28
F-test  184
furlong  7
fusion  12

Gage R&R test  22, 204
gallons  7
Gallop, George  21
Gandhi, Rajiv, National Quality Award  199–200
gas  11
gauge  15
GDT  154
Gemba  208–209
general inspection  144
generalizations  157
geometric dimensioning and tolerance  154
Gillett Method  14
Global Performance Excellence Award  200
go / no-go gauges  15
goodness of fit  82
grains  8
gram  8
graphs  2, 93
grid  19
grouping  24–40, 68, 74
guesswork  19
Guiness Breweries  160

hardness  74, 167, 159
heat  11–13
heat capacity  13
hectare  7
height  103
hemocytometer  160
high performance liquid chromatography  14
hinges  33

histogram  23, 28, 30–32
historical data  91
historical probability  45
horsepower  8, 12
House of Quality  180–181
humidity  12
hypergeometric distributions  54–56, 151
hypothesis testing  157, 162-167, 171, 175–176

imperial ton  7
improve  204–205
I-MR chart  133–134
inch  6
independent axis  34
independent events  49–50
individual data  88–89, 128, 132
individuals charts  133–135
inferential statistics  157–176
inner fence  34
inspection  137–155
integers  4
intensity  8
International Organization of Standardization  200–201
International System of Measurement  9–10
Interpolation  81
inter-quartile range  32–33
intervals  5, 24, 26, 33, 157–162
intuition  46
inventory  191
irrational numbers  4
Ishikawa diagram  182, 205
ISO  200–201

Jidoka  206
Jin  7
Joules  12
Joules per second  8
Jurtosis  88
JUSE  198
just-in-time (JIT)  206

Kaizen  200
kilogram  8
kilometers  7
kinetic energy  12
kurtosis  82–84, 160

labor 137
laser systems 154
lasers 10
Latin Square Design (LS-p) 184, 186
LCL 99–100, 126
lead time 191
lean manufacturing 207–209
Lean Six Sigma 206–209
least squares 188
length 2, 5–7, 53
leptokurtic distribution 83
life cycle 182–183
light 8–9
limiting quality level (LQL) 149
limits 24, 93–97, 123-128
line chart 38–40
linear regression 187–191
liquid 11
liquid volume 7
liters 7
LNTL 127
logical scale 13
London pound 7
long ton 7
lot size 140, 147
Lot Tolerance Percent Defective (LTPD) 149
Lottery 49–50
lower boundary 26, 33
lower control limits 94–97
lower limits 24
lower natural tolerance limits 127
lubrication 22
luminance 8-9

market analysis 179, 187–191
materials 137
mean 24, 26, 64–66, 69, 123, 172–173
mean deviation 15
measurements 1–16
 units 5–13
measures of central tendency 18, 24, 63–70
mechanical stress 22
median 24, 30, 33, 66–69, 133
merchant pound 7
mercury 12
meter 7
metric scale 6–7

metric ton 8
micrometer 1, 22
midpoint 26, 33, 69
mile 7
mode 24, 69–70, 88
mole 9–10
Motorola 202–203
moving average 187
moving range 133
Muda 208
multiple distributions 34
multiple sampling 144–147
multiple trials 49
mura 208
muri 208
mutually exclusive 47–48

nanometer 7
natural numbers 4
natural tolerance limits *see NTL*
natural variation 87, 137
negative curvilinear 36
negative skew 66–67
Newtons 8
nit 9
nominal numbers 5
non-conforming 47–49, 57, 90–91, 93, 124, 133, 140
normal distribution 38–39, 63, 76, 151, 158
not mutually exclusive 48–49
$np$-chart 97–98, 119, 122–123
NTL 127–128, 130
null hypothesis 163
numbers 3–5
numerals 4

OC curve 148–152
odds 47
off-center process 130–131
ohms 8
one event 47–49
one-tail test 163–165, 171
opacity 9
Operating Characteristic (OC) 140
ordered stem and leaf plot 30, 32
ordinate numbers 4
organizing data 17–18, 23–40

orthogonal array  184, 186
ounces  7–8
outcomes  43–44, 47, 49, 51–53
outer fence  34
outlier  34
overstocking  193

Pa•s  10
Parameter  17–18, 140
pareto chart  36–38
parts-per-million  see PPM
Pascal-second  10
Patterns  35–36, 125
p-chart  90–97, 120–122
PDCA  208
performance excellence  200
permutations  52–53
PERT  194, 205
PFMEA  205
photovoltaic cell  9
piece-to-piece variation  91
piezoelectric devices  12
pints  7
platykurtic distribution  83–84
plug gauge  15
Poisson distribution  15, 51, 59–61, 151
Poka Yoke  208
Population  17–20, 157
positive curvilinear  35
positive skew  66–67
potential energy  12
pound  7–8
Power Ball  50
PPM  1, 2, 9–10, 23, 27–30, 33, 35, 38–39, 68, 133
precision  1–3
prediction equation  188–189
prejudice  21, 139
pressure  12
probability  43, 60, 151
    classical  43–44
    relative  44–46
    subjective  46–47
    variations  47–50
process analysis  123–126
process capability  128–133
process control  126–128

process performance  132
producers  147–148
product design  179–183
program evaluation and review
    technique  194
proportion  28
prototypes  179
purity  10

QFD  180–182
qualitative measures  13
quality  22, 37, 45–46, 57, 87 104, 137
quality control  88, 90–91, 93
Quality Function Deployment  180–182
quantitative measures  13
quartiles  32
quarts  7

R charts  104–113, 119, 122–123, 125, 133
R&D  179
radiation  22
Randomized Block Design (RB-p)  184–185
randomness  19, 23, 138, 144
range  23–24, 27, 71, 133
rational numbers  4
rational subgroup  91
ratios  5
raw data  23
reaming  35–37, 39–40
reflection  9
reflectivity  9
refraction  9
refractivity  9
regression  38
rejectable Quality Level (RQL)  149
relative cumulative histogram  31
relative frequency chart  28
relative probability  44–46
relativity  28
reliability  137, 182–183
reorder points (ROP)  191–192
repeatability  22
reproducibility  22
research and development  179
resistance  8, 22
Rice's rule  28
rivets  17–18

robotics 154
rotors 47–49
roughness 35, 39–40
rounding 64–65
RQL 149
rule violations 125
ruler 13
run chart 133–135

S charts 104, 113–117, 119
safety stock 191
samples 17, 53, 88, 157
　　size 18–19, 119
sampling 18–23, 137–155
　　double sampling 144–147
　　multiple sampling 144–147
　　single sampling 140–144, 148–151
scale 1–3, 5, 6, 13–15
scatter plots 34–36, 38
scheduling 191
Scott's rule 28
Scoville scale 14
sequential sampling 147
serial numbers 5, 138
service level (SL) 191
shaft diameters 133–135
shear pins 22
Shewhart, William 87
short ton 8
SI 9–10
Siemens 8
sigma zones 124, 128
single period models (SPM) 191, 193–194
Six Sigma 202–209
size 18–19, 140, 142
skewness 66–67, 69, 82, 88
skip lot sampling 147
SL 127
SMED 208
societies 200–202
SPC 14, 87–101, 103–117, 137
special cause variation 87
specification limit 127
specifications 126–128
spectrometer 10
split stem and leaf plot 30
square measures 7

stamping 37–38
standard deviation 15, 24, 71–75, 88, 113–117, 123–124, 132
standardization 79
standardized scales 5, 7
standardized tables 140–141
standards 200–202
statistical process control *see SPC*
steel rods 5–6
stem and leaf plot 29–33
stratification 93, 138-139
stress 22
Sturgis' rule 28
subcomponents 137
subgroups 88–89, 119-123, 128, 135
subjective probability 46–47
subjective scale 13–14
sufficiency 140
sum of squares 174
surface 35, 39, 104
surface variability 2–3
system design 179–183
system operation 191–194

t distribution 160
t score 163
t test 157, 184
T.S. 163, 168–169, 171
tabulation 23–24
taguchi 186
tally 24–25
tape measure 2, 13
target values 87
technicians 180
temperature 2, 5, 10–12, 22
Tesla 8
test statistic *see T.S.*
testing 141–142, 157
thermal stress 22
thermistors 12
thermocouples 12
thermometer 12
time-to-time variation 91
tolerance 126–128
ton 7–8
tone 8
total productive maintenance 208

total quality management  198
total sum of squares  174, 184
tower pound  7
Toyoda, Eiji  206
TPM  208
TQM  198–200, 202, 206
transformer  8
translucency  2, 9
transparency  9
trends  36, 38
trial  47, 49
troy pound  7
tungsten inert gas (TIG)  13
turbine  12
two populations  169–172
two-tail test  164–165, 168
Type I error  123–124, 165–167
Type II error  124, 165–167

u-charts  91, 100–101
UCL  94–97, 99–100, 121–122, 126
Understocking  193
units of measure  5–13
UNTL  127–128, 131
upper boundary  26, 33
upper control limits  *see UCL*
upper limits  24
upper natural tolerance limits  *see UNTL*
upward pattern  125

validity  19
value stream map  207
vaporization  12
variability  18–19
variable charts  45, 132
variables  14, 18, 21–23, 53, 87, 119–123

variance  2, 19, 71–75, 172–176
variations  47–50, 87, 91
verify  205
Vernier scale  1
viscometers  10-11
viscosity  10-11
vision systems  154–155
voice of the customer  180, 207
voltage  2, 8, 12, 53
volume  2, 7

wattage  8, 12
wavelength  9
wear out time (wot)  183
weight  2, 5, 7–8, 53, 103
weighted mean  65
weighted moving average  187
whisker plots  33–35
whole numbers  4
within-piece variation  91
working conditions  22
worn dies  37

X-bar charts  104–113, 119, 122–123, 125–126, 133

yard  6, 13

z distribution  78–81, 157–158
z score  79–81, 130, 157–158, 163
zero  4–5
zone rules  124

$\mu$  157–158, 167–169
$\sigma$  157, 159–162, 167–169